Introduction to Proteomics

Introduction to Proteomics

Kate Banks

Larsen & Keller
www.larsen-keller.com

Introduction to Proteomics
Kate Banks
ISBN: 978-1-64172-611-5 (Hardback)

⊟ Larsen & Keller

Published by Larsen and Keller Education,
5 Penn Plaza,
19th Floor,
New York, NY 10001, USA

Cataloging-in-Publication Data

Introduction to proteomics / Kate Banks.
    p. cm.
Includes bibliographical references and index.
ISBN 978-1-64172-611-5
1. Proteomics. 2. Proteins. 3. Molecular biology. I. Banks, Kate.
QP551 .I58 2022
574.192 45--dc23

For more information regarding Larsen and Keller Education and its products, please visit the publisher's website www.larsen-keller.com

TABLE OF CONTENTS

PREFACE

This book is a culmination of my many years of practice in this field. I attribute the success of this book to my support group. I would like to thank my parents who have showered me with unconditional love and support and my peers and professors for their constant guidance.

The scientific study of proteins is known as proteomics. Proteins perform various essential functions in living organisms. Proteomes are the sets of proteins that are modified and created by an organism. It is useful in identification of the increasing number of proteins. Proteomics utilizes the genetic information provided by various genome projects such as the human genome project. It explores the proteomes from several levels of protein composition, structure and activity. Proteomics is a vital component of functional genomics which attempts to explain the functions and interactions of genes. Some of the processes studied within this discipline are protein exploration, protein interactions, protein phosphorylation, etc. The topics included in this book on proteomics are of utmost significance and bound to provide incredible insights to readers. Such selected concepts that redefine this field have been presented herein. Through this book, we attempt to further enlighten the readers about the new concepts in this field.

The details of chapters are provided below for a progressive learning:

Chapter – What is Proteomics?

The entire set of proteins which can be expressed by a genome, tissue, cell or organism at a certain time is known as a proteome. The large-scale study of proteins is referred to as proteomics. This chapter has been carefully written to provide an easy understanding of the various aspects of proteomics.

Chapter – Protein: Types and Processes

The large biomolecules or macromolecules composed of one or more long chains of amino acid which are connected through a covalent peptide bond are known as proteins. They are classified into different types such as blood proteins, globular protein, GST-tagged proteins and His-tagged proteins. These diverse classifications of proteins have been thoroughly discussed in this chapter.

Chapter – Branches of Proteomics

There are various branches of proteomics such as structural proteomics, interaction proteomics, phosphoproteomics, immunoproteomicsm, secretomics, activity-based proteomics, neuroproteomics and quantitative proteomics. This chapter closely examines these branches of proteomics to provide an extensive understanding of the subject.

Chapter – Protein Methods

The techniques used to study proteins are known as protein methods. Some of these methods are protein purification, protein identification, protein mass spectrometry, protein sequencing, two-dimensional gel electrophoresis, threading, etc. The topics elaborated in this chapter will help in gaining a better perspective about these protein methods.

Chapter – Protein-Protein Interaction

The physical contacts of high specificity established between two or more protein molecules are referred to as protein–protein interactions. The main focus areas of protein-protein interaction are protein–protein interaction prediction and protein–protein interaction screening. This chapter has been carefully written to provide an easy understanding of these aspects associated with protein-protein interaction.

Kate Banks

What is Proteomics?

The entire set of proteins which can be expressed by a genome, tissue, cell or organism at a certain time is known as a proteome. The large-scale study of proteins is referred to as proteomics. This chapter has been carefully written to provide an easy understanding of the various aspects of proteomics.

PROTEOME

The proteome is defined as the entire set or complement of proteins that is or can be expressed by a cell, tissue, or organism. The term originates from a word play blending the words protein and genome to create a newer term now called "The Proteome". Since the set of expressed proteins in a cell, tissue or organelle is dynamic the actual proteome reflects a particular set of biochemical conditions viewed as a snapshot-in-time of the biochemical system studied. In addition, the number of proteins, for example, in the human proteome, can be as large as 2 million. Marc Wilkins is reported to have coined the term proteome in 1994 in a symposium on "2D Electrophoresis: from protein maps to genomes" held in Siena, Italy and part of his PhD thesis appeared in print in 1996. The publication showed that proteins can be separated and identified by two-dimensional (2-D) electrophoresis allowing for protein-based gene expression analysis. Identification of unknown proteins was achieved by matching their amino acid composition, estimated pI and molecular weight against all E. coli entries in the SWISS-PROT database.

This approach allowed large-scale screening of the protein complement of simple organisms, or tissues in normal and disease states. In the following years proteomics became a popular approach for the study and identification of whole protein assemblies leading to a new biochemistry branch called "proteomics" that is concerned with the analysis of the structure and functions of proteins occurring in living organisms. The term proteome was used by Mark to describe the entire complement of proteins expressed by a genome, cell, tissue or an organism. Since its appearance, it has been applied to several different types of biological systems. For example, a cellular proteome describes the whole collection of proteins found in a particular cell type under a particular set of environmental conditions such as exposure to hormone stimulation or different drugs. However the term "complete proteome" may be used to describe the complete set of proteins from all of the various cellular proteomes which could be very used to roughly describe the whole protein equivalent of the genome. Furthermore the term "proteome" has also been used to refer to the collection of proteins in certain sub-cellular biological systems. For example, all of the proteins in a virus can be called a viral proteome. In conclusion, proteomes are now routinely investigated using a combined approach of amino acid analysis and peptide-mass fingerprinting allowing gene products to be linked to homologous genes in a variety of organisms.

PROTEOMICS

The term proteomics was coined in mid 1990s at the back drop of successful genomics. In bioinformatics point of view proteomics is the databases of protein sequence, databases of predicted protein structures and more recently, databases of protein expression analysis. As more protein structures are identified, the relationship between structure and functions became easier to predict.

In addition, databases of protein structure and corporating tools facilitating the identification of common protein structure and their predicted functions. In this technique individually purified ligands such as proteins, peptides, antibodies, antigens, and carbohydrates are spotted on to a derivatized surface and are generally used for examining protein expression levels for protein profiling.

A major challenge facing plant biotechnology and other bioinformatics research community is the translation of complete genome DNA sequence data into protein structure and predicted functions. Such a steps will provide the key link between the genotypes of an organism and its expressed phenotype.

The growth of proteomics is a direct result of advances made in large scale nucleotide sequencing of expressed sequence tags (EST). Although mass spectrometry or more popularly MS technology has been considered as versatile tool for examining simultaneous expression of more than 1000 proteins and identification, mapping of post-translational modifications.

These methods performed in a latest array of technology resulted in large-scale characterization of protein location, protein-protein interaction and protein functions:

Method	Description	Applications
• Mass spectrophoto-meter	Digest protein and fragment peptide to identification proteins,	Protein identification, sequence post translational modification
• Chip	Synthesise proteins, peptides, antigens, antibody into a every format and spot onto slides.	Protein interaction with protein, lipid and small molecules, drug discover, post translational modi-fications.
• Bioinformatics	Insilico proteomics.	Mining database predicting protein interaction.

Insilico methodologies are being developed to identify protein interaction from genome sequence. For example, 6809 putative protein-protein interaction has been identified in Escherichia coli and more than 45,000 have been identified in yeast and large number of these interactions is functionally related.

Types of Proteomics

Structural Proteomics

One of the main targets of proteomics investigation is to map the structure of protein complexes or the proteins present in a specific cellular organelle known as cell map or structural proteins. Structural proteomics attempt to identify all the proteins within a protein complex and characterization all protein-protein interactions. Isolation of specific protein complex by purification can simplify the proteomic analysis.

Functional Proteomics

It mainly includes isolation of protein complexes or the use of protein ligands to isolate specific types of proteins. It allows selected groups of proteins to be studied its chracteristics which can provide important information about protein signalling and disease mechanism etc.

Significance of Proteomics

Protein Profiling

Bioinformatics has been widely employed in protein-profiling, where question of protein structural information for the purpose of protein identification, characterization and database is carried out. The spectrum of protein expressed in a cell type provides the cell with its unique identity. It explores how the protein complement changes in a cell type during development in response to environmental stress.

Protein Arrays

Protein microarrays facilitate the detection of protein protein interaction and protein expression profiling. Several protein microarray examples indicate that protein arrays hold great promise for the global analysis of protein-protein and protein-ligand interaction.

Proteomics to a Phosphorylation

In post-translational modification of protein, mass spectrometer (MS) can be used to identify novel phosphorylation. Measure changes in phosphorylation state of protein takes place in response to an effective and determining phosphorylation sites in proteins.

Identification of phosphorylation sites can provide information about the mechanism of enzyme regulation and protein kinase and phosphotases involved. A proteomics approach for this process has an advantage that one can study all the phosphorylating proteins in a cell at the same time.

Proteome Mining

Proteome mining is a functional proteomic approach used to extract information from the analysis of specific sub-proteomics. In principle, it is based on the assumption. In principle, it is based oil the assumption that all drug like molecule selectively compete with a natural cellular ligand for a binding site on a protein target.

PROTEOMICS TECHNOLOGIES AND THEIR APPLICATIONS

The dynamic role of molecules to support the life is documented since the initial stages of biological research. The "proteome" can be defined as the overall protein content of a cell that is characterized with regard to their localization, interactions, post-translational modifications and turnover, at a particular time. The term "proteomics" was first used by Marc Wilkins in 1996 to denote the "PROTein complement of a genOME". Most of the functional information of genes is characterized by the proteome. The proteome of eukaryotic cells is relatively complex and exhibits extensive dynamic range. Moreover, prokaryotic proteins are responsible for pathogenic mechanisms; however, their analysis is challenging due to huge diversity in properties such as dynamic range in quantity, molecular size, hydrophobicity and hydrophilicity.

Proteomics is crucial for early disease diagnosis, prognosis and to monitor the disease development. Furthermore, it also has a vital role in drug development as target molecules. Proteomics is the characterization of proteome, including expression, structure, functions, interactions and modifications of proteins at any stage. The proteome also fluctuates from time to time, cell to cell and in response to external stimuli. Proteomics in eukaryotic cells is complex due to post-translational modifications, which arise at different sites by numerous ways.

Proteomics is one of the most significant methodology to comprehend the gene function although, it is much more complex compared with genomic. Fluctuations in gene expression level can be determined by analysis of transcriptome or proteome to discriminate between two biological states of the cell. Microarray chips have been developed for large-scale analysis of whole transcriptome. However, increase synthesis of mRNA cannot measure directly by microarray. Proteins are effectors of biological function and their levels are not only dependent on corresponding mRNA levels but also on host translational control and regulation. Thus, the proteomics would be considered as the most relevant data set to characterize a biological system.

The conventional techniques for purification of proteins are chromatography based such as ion exchange chromatography (IEC), size exclusion chromatography (SEC) and affinity chromatography. For analysis of selective proteins, enzyme-linked immunosorbent assay (ELISA) and western blotting can be used. These techniques may be restricted to analysis of few individual proteins but also incapable to define protein expression level. Sodium dodecyl sulfate-polyacrylamide gel electrophoresis (SDS-PAGE), two-dimensional gel electrophoresis (2-DE) and two-dimensional differential gel electrophoresis (2D-DIGE) techniques are used for separation of complex protein samples.

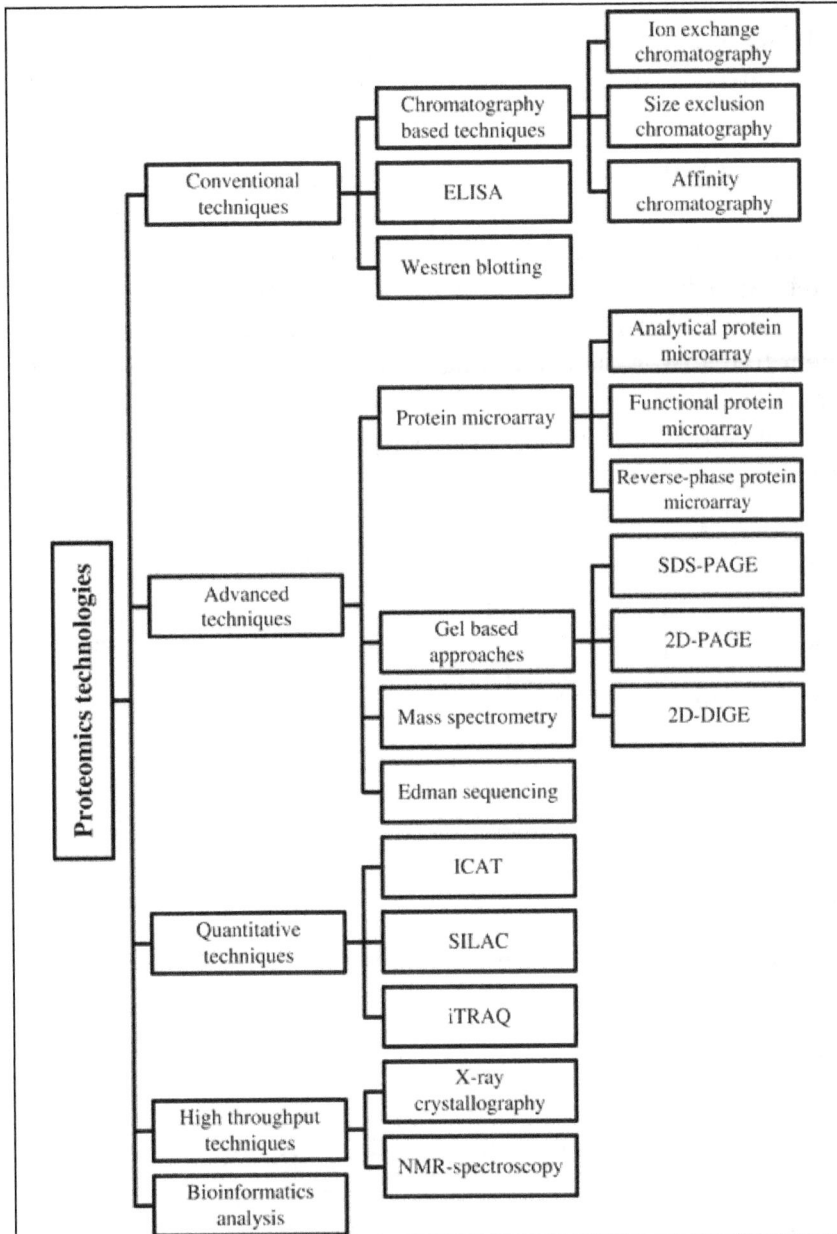

An overview of proteomics techniques.

Protein microarrays or chips have been established for high-throughput and rapid expression analysis; however, progress of a protein microarray enough to explore the function of a complete genome is challenging. The diverse proteomics approaches such as mass spectrometry (MS) have developed to analyze the complex protein mixtures with higher sensitivity. Additionally, Edman degradation has been developed to determine the amino-acid sequence of a particular protein. Isotope-coded affinity tag (ICAT) labeling, stable isotope labeling with amino acids in cell culture (SILAC) and isobaric tag for relative and absolute quantitation (iTRAQ) techniques have recently developed for quantitative proteomic. X-ray crystallography and nuclear magnetic resonance (NMR) spectroscopy are two major high-throughput techniques that provide three-dimensional (3D) structure of protein that might be helpful to understand its biological function.

With the support of high-throughput technologies, a huge volume of proteomics data is collected. Bioinformatics databases are established to handle enormous quantity of data and its storage. Various bioinformatics tools are developed for 3D structure prediction, protein domain and motif analysis, rapid analysis of protein–protein interaction and data analysis of MS. The alignment tools are helpful for sequence and structure alignment to discover the evolutionary relationship. Proteome analysis provides the complete depiction of structural and functional information of cell as well as the response mechanism of cell against various types of stress and drugs using single or multiple proteomics techniques.

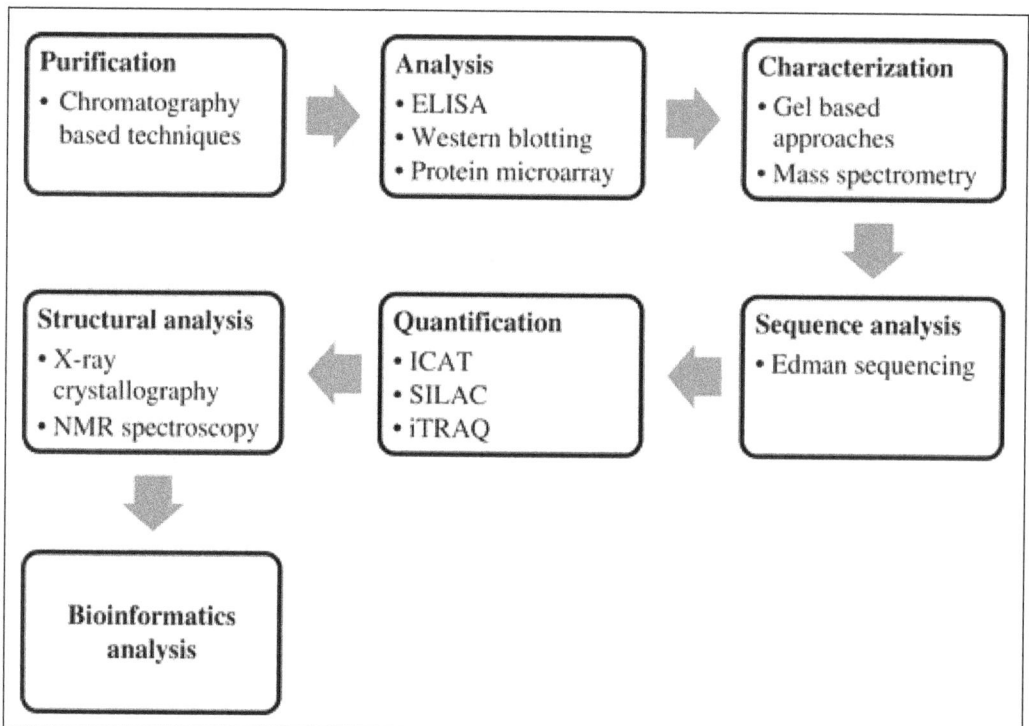

Applications of proteomics techniques.

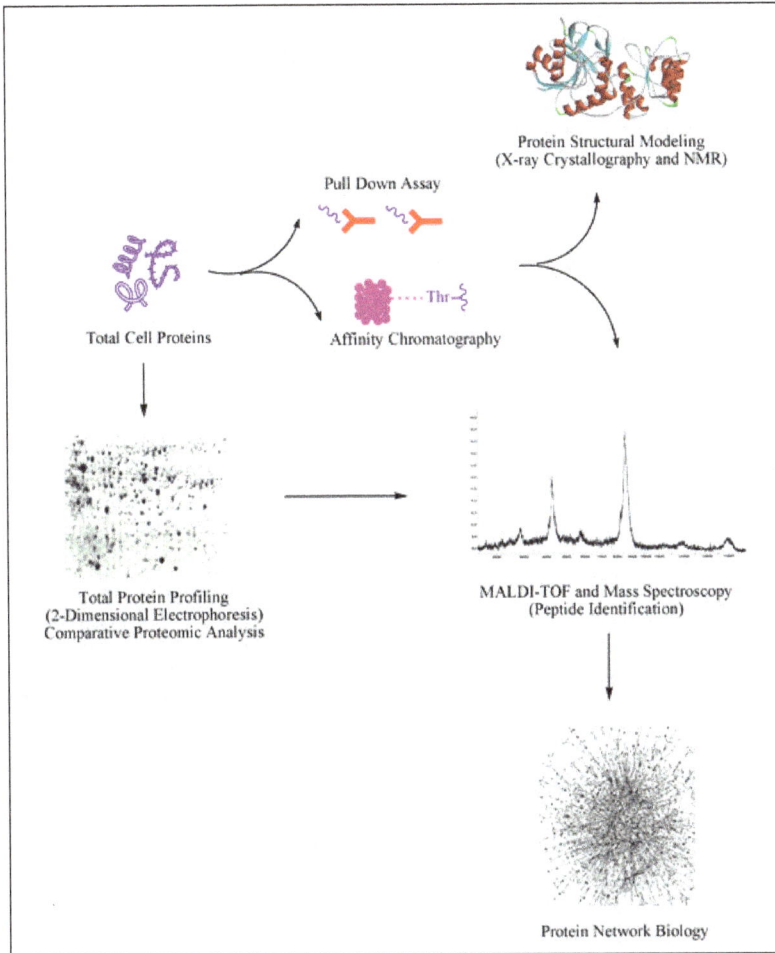

Schematic representation of protein analysis. The conventional methodology for protein analysis includes protein extraction, purification and structural studies. Cells or tissue are processed by various physical (sonication) and chemical (detergents) techniques for the extraction of total protein. Based upon physiochemical nature of polypeptides, the protein of interest can be separated out by different chromatographic techniques. Various methods including X-ray crystallography, NMR and MALDI-TOF are extensively used for structural elucidation and functional characterization of proteins. Nowadays, high-throughput techniques including total proteome analysis and MALDI-TOF are employed to study protein network biology.

Chromatography-based Techniques

Ion Exchange Chromatography

The IEC is a versatile tool for the purification of proteins on the basis of charged groups on its surface. The proteins vary from each other in their amino-acid sequence; certain amino acids are anionic while others are cationic. The net charged contain by a protein at physiological pH is evaluated by equilibrium between these charges. Initially, it separates the protein on the basis of their charge nature (anionic and cationic), further on the basis of comparative charge strength. The IEC is highly valuable due to its low cost and its capacity to persist in buffer conditions.

A most important virulence factor of Helicobacter pylori is the Neutrophil Activator Protein (HP-NAP) that is able to activate human neutrophil by secreting mediators and reactive oxygen species. The HP-NAP is a potential diagnostic marker for H. pylori and as well a probable drug target and vaccine candidate. One step anionic exchange chromatography has been designated by Shih et al. to purify the recombinant HP-NAP expressed in B. subtilis with 91% recovery. The mussel adhesive proteins (MAPs) have distinctive biocompatible and adhesive properties that are useful for biomedical and tissue engineering. Choi et al. expressed the recombinant MAPs in E. coli and successfully purified through IEC. Antifungal proteins from B. subtilis strain B29 were purified through IEC on diethylaminoethyl.

Nigella sativa proteins that retain immune modulatory action have been fractionated through IEC and four peaks were received in complete fractionation. Proteins expressed in transgenic plants commercially values in pharmaceutical products. An example is Aprotinin; an inhibitor of serine proteases that were expressed in corn seed and purified. Cysteine proteases are the key mediators of mammalian apoptosis and inflammation that are expressed in E. coli and purified by Garica-calvo et al. for better understanding of catalytic properties. The serum consists of various chemokines, cytokines, peptide hormones and proteolytic fragments of large proteins that can be purified using strong cation exchange chromatography.

Size Exclusion Chromatography

SEC separates the proteins through a porous carrier matrix with distinct pore size on the basis of permeation; therefore, the proteins are separated on the basis of molecular size. The SEC is robust technique capable of handling proteins in diverse physiological conditions in the presence of detergents, ions and co-factors or at various temperatures. The SEC is used to separate low molecular weight proteins and is a powerful tool for purification of non-covalent multi-meric protein complexes under biological conditions.

The soluble factors produced by Trichomonas vaginalis have the ability to damage the target cells and involved in pathogenesis of trichomoniasis. The phospholipase A2-like lytic factor has been purified and further characterization exhibited 168 and 144 kDa two fractions. The antimicrobial peptides synthesized by marine bacterium Pseodoalteromonas have been purified from culture supernatant through SEC that possess strong inhibitory effect against pathogens involved in skin infections. Cytosolic proteins of Arabidopsis thaliana have been purified to understand how cell coordinates diverse mechanical, metabolic and developmental activities. Purification of intrinsically disordered proteins of A. thaliana was also carried out through SEC. These are expressed during advanced stage of seed development and have a significant role in transcription regulation and signal transduction.

Affinity Chromatography

The affinity chromatography was a major breakthrough in protein purification that enables the researcher to explore protein degradation, post-translational modifications and protein–protein interaction. The basic principle behind the affinity chromatography is the reversible interaction between the affinity ligand of chromatographic matrix and the proteins to be purified.

The affinity chromatography has a wide range of applications in identification of microbial enzymes principally involved in the pathogenesis. Homodimer and heterodimer of HIV-I reverse transcriptase were rapidly purified by metal chelate affinity chromatography. The practical applications of bacteriophages in field of biotechnology and medicine persuade excessive requirement of the phage purification. The T4 bacteriophages have been purified from bacterial debris and other contaminating bacteriophages. The bacterial cells in 'competitive phage display' produced both fusion protein and wild-type proteins. The fusion proteins were integrated into phage capsid and permitted the effective purification of T4 bacteriophages.

A group of amyloid binding proteins interact with different forms of amyloidogenic protein and peptides, therefore modify their pathological and physical role. Affinity chromatography is potentially applied for the diagnosis of Alzheimer's disease by purification of Alzheimer's amyloid peptide from human plasma. The immobilized metal ion affinity chromatography purified the heterologous proteins comprising zinc finger domains. Hexa-histidine affinity tags displayed different affinities to the immobilized metal ions even though both contain same type of domain. However, zinc finger proteins vary in biochemical properties.

Plasma proteins such as factor IX, factor XI, factor VIII, antithrombin III and protein C have been purified through affinity chromatography at industrial scale for therapeutic use. Various ligands have been purified and applied in purification of antibodies. The examples include the lectins for IgM and IgA purification whereas proteins A and G for the purification of IgG molecules.

Enzyme-linked Immunosorbent Assay

In 1971, Engvall and Pearlmann published the first paper on ELISA and quantified the IgG in rabbit serum using the enzyme alkaline phosphatase. The ELISA is highly sensitive immunoassay and widely used for diagnostic purpose. The assay utilizes the antigen or antibodies on the solid surface and addition of enzyme-conjugated antibodies to and measure the fluctuations in enzyme activities that are proportional to antibody and antigen concentration in the biological specimen.

The diagnosis of paratuberculosis or John's disease was made possible by Ethanol Vortex ELISA. The assay distinguished the surface antigens of Mycobacterium avium subspecies paratuberculosis. Capture ELISA was established for detection of Echinostoma caproni in experimentally infected rats. This assay was based on recognition of excretory–secretory antigens by polyclonal rabbit antibodies. The detection limit was 60 ng/ml in fecal sample and 3 ng/ml in sample buffer. Deoxynivalenol (DON), a powerful mycotoxin produced by Fusarium graminearum is a major contaminant of barley and wheat and leads to Fusarium Head Blight. Indirect competitive ELISA for the identification of DON in wheat was developed with detection limit between 0.01 and 100 µg/mL in grains.

Wheat proteins causes allergic reactions in susceptible individuals that have been traced in foods to protect wheat-sensitive individuals using commercially available ELISA kits. Sandwich ELISA was used for the detection of Cry1Ac protein of Bacillus thuringiensis from transgenic BT cotton as their release adversely affect the environment. Indirect competitive ELISA was developed to detect Botrytis cinerea in tissues of fruits. B. cinerea is a phyopathogenic fungus responsible for gray mold and often present as latent infection and deteriorate the healthy fruits. Digital ELISA is capable of detecting single molecule in the blood. The assay was able to detect prostate-specific antigen (PSA) in the serum at low concentration of 14 fg/ml. This assay was capable to detect 1,1-Dichloro-2,2-bis (p-chlorophenyl) ethylene (p,p'-DDE); a metabolite of insecticide and persistent organic pollutant that accumulates in food chain and environment.

Western Blotting

Western blotting is an important and powerful technique for detection of low abundance proteins that involve the separation of proteins using electrophoresis, transfer onto nitrocellulose membrane and the precise detection of a target protein by enzyme-conjugated antibodies. Western blotting is a dominant tool for antigen detection from various microorganisms and is quite helpful in diagnosis of infectious diseases. The seroprevalence of Herpes Simplex Virus type 2 (HSV-2) in African countries was investigated by measuring the specific immunoglobulin G in the sera of patients. Leishmania donovani is responsible for visceral leishmaniasis, which is classically diagnosed by the presence of Hsp83 and Hsp70 antigens in the bone marrow, spleen and liver.

Western blotting was carried out by Li et al. for identification and validation of 10 rice reference proteins. Elongation factor 1-α and heat-shock proteins were the most expressed proteins in rice. Kollerova et al. identified the Plum Pox Virus (PPV) capsid proteins from infected Nicotiana benthamiana. The expression of PfCP-2.9 gene of Plasmodium falciparum in tomato was confirmed through western blot analysis. Specific IgE against Ara h1, Ara h2 and Ara h3 was determined in peanut allergic patients through western blotting.

Edman Sequencing

Edman sequencing was developed by Pehr Edman in 1950 to determine the amino-acid sequence in peptides or proteins. The method comprises chemical reactions that eliminate and identify amino acids residue that is present at the N-terminus of polypeptide chain. Edman sequencing played a major role in development of therapeutic proteins and quality assurance of biopharmaceuticals.

Brucella suis survive and replicate in macrophage due to the acidification. The proteins that are involved in this acidification were identified. Edman degradation and comparison of 13 N-terminal amino-acid sequences revealed that these were signal peptides for its periplasmic location. The protein in B. suis that was involved in membrane permeability at acidic environment was Omp25. The causative agent of hemorrhagic fever, Lassa virus belongs to family of Arenaviridae. The Lassa virus synthesis glycoproteins which are cleaved into GP-1 (amino-terminal subunit) and GP-2 (Carboxy-terminal subunit) after translation and are primarily involved in pathogenesis. The Edman degradation analysis of GP-2 revealed N-terminal tripeptide GTF262.

The prevalence of sesame seed allergy has been increasing due to the use of bakery products and fast-food. The major allergic proteins of Sesamum indicum have been identified from allergic patients through 2D-PAGE and SDS-PAGE and then further analyzed through Edman sequencing. IgE binding epitopes of these proteins were identified that might be helpful in immunotherapeutic approaches. The proteins from leaf sheaths of rice were extracted and analyzed through MS and Edman sequencing to determine its function. The amino-acid sequence of majority of proteins analyzed by both techniques have similar results, therefore suggesting the use of these techniques in combination for the identification of plant protein.

Advanced Techniques

Protein Microarray

Protein microarrays also known as protein chips are the emerging class of proteomics techniques capable of high-throughput detection from small amount of sample. Protein microarrays can be classified into three categories; analytical protein microarray, functional protein microarray and reverse-phase protein microarray.

Analytical Protein Microarray

Antibody microarray is the most representative class of analytical protein microarray. After antibody capture, proteins are detected by direct protein labeling. These are typically used to measure the expression level and binding affinities of proteins. High-throughput proteome analysis of cancer cells was carried out through antibody microarray for differential protein expression in tissues derived from squamous carcinoma cells of oral cavity. Antibody array was also used for protein profiling of bladder cancer. Microarray immunoassay was used for detection of Staphylococcal enterotoxin B, cholera toxin, Bacillus globigii and B. ricin. Analytical and experimental approaches have been developed for identification of cellular signaling pathways and to characterize the plant kinases through protein microarray. Mitogen-activated protein kinases (MAPKs) from Arabidopsis have been characterized. MAPKs are highly conserved single transduction and universal molecules in plants that respond to wide range of extracellular stimuli.

Functional Protein Microarray

Functional protein microarray is constructed by means of purified protein, thus permits the study of various interactions including protein–DNA, protein–RNA and protein–protein, protein–drug, protein–lipid, enzyme–substrate relationship. The first use of functional protein microarray was to analyze the substrate specificity of protein kinases in yeast. Functional protein microarray characterized the functions of thousands of proteins. The protein–protein interaction of A. thaliana was studied and Calmodulin-like proteins (CML) and substrates of Calmodulin (CaM) were identified.

Reverse-phase Protein Microarray

Cell lysates obtained from different cell states are arrayed on nitrocellulose slide that are probed with antibodies against target proteins. Afterwards, antibodies are detected with fluorescent, chemiluminescent and colorimetric assays. For protein quantification, reference peptides are printed on slides. These microarrays are used to determine the altered or dysfunction protein indicative of a certain disease. The analysis of hematopoietic stem cell and primary leukemia samples through reverse-phase protein microarray was found to be highly reproducible and reliable for large-scale analysis of phosphorylation state and protein expression in human stem cells and acute myelogenous leukemia cells. Reverse-phase protein microarray approach was evaluated for quantitative analysis of phosphoproteins and other cancer-related proteins in non-small cell lung cancer (NSCLC) cell lines by monitoring the apoptosis, DNA damage, cell-cycle control and signaling pathways.

Gel-based Approaches

Sodium Dodecyl Sulfate-polyacrylamide Gel Electrophoresis

SDS-PAGE is a high resolving technique for the separation of proteins according to their size, thus facilitates the approximation of molecular weight. Proteins are capable of moving with electric field in a medium having a pH dissimilar from their isoelectric point. Different proteins in mixture migrate with different velocities according to the ratio between its charge and mass. However, addition of sodium dodecyl sulfate denatures the proteins, therefore separate them absolutely according to molecular weight.

The protein profiling of Mycoplasma bovis and Mycoplasma agalactiae through SDS-PAGE has high diagnostic value as these species are difficult to differentiate with routine diagnostic procedures. The outer membrane proteins from E. coli strains in which ability to form K1 antigen is absent were analyzed through SDS-PAGE. It exhibited varied degree of susceptibility to the human serum. Extracellular protein profile of Staphylococcus spp. was also constructed and their characterization was achieved. The antigenic proteins of Streptococcus agalactiae have been characterized to test the immunogenicity of mastitis vaccine.

The cleome spp. are consumed as green vegetables in African countries and highly valuable for the treatment of cough, fever, asthma, rheumatism and many other diseases. The comparative analysis of leaf and seed proteins of cleome spp. was carried out by SDS-PAGE. The profiling of seed and leaf storage proteins of chickpea (Cicer arientinum) was conducted under drought stress and non-stress conditions. The seed storage proteins of Brassica species are also identified to evaluate the genetic divergence in different genotypes. The influence of heat treatment and addition of demineralized whey on the soluble protein composition of the skim milk was investigated. High molecular weight complexes were formed during the addition of demineralized whey as well as heat treatment which was determined by SDS-PAGE. Large-scale production of insulin is helpful for the management of diabetes, therefore different approaches and species have been used for the production of insulin. Elamin et al. purified and characterized the pancreatic insulin from the Camelus dromedaries.

Two-dimensional Gel Electrophoresis

The two-dimensional polyacrylamide gel electrophoreses (2D-PAGE) is an efficient and reliable method for separation of proteins on the basis of their mass and charge. 2D-PAGE is capable of resolving ~5,000 different proteins successively, depending on the size of gel. The proteins are separated by charge in the first dimension while in second dimension separated on the basis of differences between their mass. The 2-DE is successfully applied for the characterization of post-translational modifications, mutant proteins and evaluation of metabolic pathways. Neidhardt and van Bogelen introduced the highly sensitive technique of 2-DE into the bacterial physiology.

The membrane proteins from the cell wall of Listeria innocua and Listeria monocytogenes involved in the host–pathogen interactions were analyzed with 2-DE and 30 different proteins of two strains were identified. This approach was useful for the comparative study of exotoxins and virulence factors released by enterotoxigenic strains of two food-derived Staphylococcus aureus strains. Pseudomonas aeruginosa secrete numerous proteins during different stages of infection as seen in isolates obtained from cystic fibrosis patients. Current improvements in the 2D-PAGE have been used to study the metabolic system of B. subtilis and a PyrR bacterial regulatory protein was characterized.

Large number of proteins were detected during the seed development in Ocotea catharinensis, and profile was constructed by characterizing these proteins during each developmental stage. Protein extraction from grapes is challenging due to the low concentration of proteins, high activity of proteases and high level of interfering compounds such as polyphenols, flavonoids, terpenes, lignans and tannins; however, Marsoni et al. successfully extracted the proteins from grape tissue through 2-DE. Islam et al. also extracted the proteins from mature rice leaves and applied in the proteome analysis.

Two-dimensional Differential Gel Electrophoresis

2D-DIGE utilizes the proteins labeled with CyDye that can be easily visualized by exciting the dye at a specific wavelength. Cell wall proteins (CWPs) of toxic dinoflagellates Alexandrium catenella labeled with Cy3 have been identified through 2D-DIGE. Quantitative analysis of Brucella suis proteins has been carried out under long-term nutrient starvation and ~30 proteins were identified that vary in concentration among bacteria grown at stationary phase in medium with different nutrient levels. About 70% of regulated proteins showed an increase in expression. The proteins are also involved in regulation, adaptation to harsh condition and transportation. The characterization of proteins expressed in rat neurons have been carried to understand the pathogenesis of West Nile virus.

The plasma membrane responds to the biotic and abiotic stress in plants, therefore the characterization of plasma proteins provides new perception about the plant-specific biological functions. Komatsu characterized the plasma membrane proteome of rice and A. thaliana. The role of apoplastic proteins of 10-day-old rice plants in salt stress response was investigated. For differential analysis, soluble apoplastic proteins from rice shoot stem were extracted and compared with untreated and were found to be involved in oxidation-reduction reaction, carbohydrate metabolism and protein degradation and processing. During ovule development of Pinus tabuliformis, female gametophyte cellularization is a vital process regulated by multiple proteins, which were first extracted in anaphase and prophase then separated through 2D-DIGE.

The biological drugs produced during cell culture technology constitute host cell proteins (HCP) as most important group of impurities. The HCP has diverse molecular and immunological properties and should be effectively monitored and removed during downstream processing. 2D-DIGE was used to screen the HCP composition in CHO cell culture and to compare HCP difference between null cell culture and monoclonal antibody producing cells. The quantitative changes in red blood cell membrane proteins in sickle cell disease were analyzed and the contents of 49 gel spots were found altered by 2.5-fold in comparison with normal cells.

The 2-DE remains a method of choice in proteomic research, though certain limitations enervate its potential as a principal separation technique in modern proteomics. Therefore, the state of the art instrumentation and techniques are rapid expanding as a new means of gel-free analytical techniques. The advancement of MS coupled with shotgun proteomics can find newer directions for sensitive and high quantity protein profiling with more accurate quantification. The chemical label-based approaches remained popular in quantitative proteomics, these methods also have certain drawbacks. The quantitative plant proteomics is more challenging due to problems associated with protein extraction, abundance of proteins in some plants tissues and the lack of well-marked genome sequences. The higher resolution power of MS, exact mass measurements, higher scanning rates and precise chromatogram alignment are essential feature for the successful use of MS in proteomics.

Quantitative Techniques

ICAT Labeling

The ICAT is an isotopic labeling method in which chemical labeling reagents are used for quantification of proteins. The ICAT has also expanded the range of proteins that can be analyzed and permits the accurate quantification and sequence identification of proteins from complex mixtures. The ICAT reagents comprise affinity tag for isolation of labeled peptides, isotopically coded linker and reactive group.

Mycobacterium tuberculosis is considered as a most important human pathogen that contain ~4,000 genes. The proteome analysis was carried out using a combination of Liquid Chromatography (LC), Tandem Mass Spectrometry (MS/MS) and ICAT. The combination of techniques offers comprehensive understanding of biological system and provides additional information. The systemic proteome quantification was carried out possible through ICAT during cell cycle of Saccharomyces cerevisiae that supported the cognition of gene functions. The levels of reactive nitrogen species and reactive oxygen species increase in living cells during abiotic and biotic stress.

The reversible oxidation of protein residues may assist as redox sensors and signal transducers for transmission of anti-stress responses. The thiol group on cysteine residue is sensitive to oxidative species and upon oxidation can modulate protein function. ICAT reagents precisely react with thiol group of cysteine residues, therefore the technique coupled with MS is useful to quantify the thiol-containing redox proteins. The tumor-specific proteins were analyzed through ICAT and MS from the aspirated fluid of breast tumor patients at earlier stages. Beta-globin, hemopexin, lipophilin B and vitamin D-binding proteins were overexpressed while Alpha2HS-glycoprotein was under expressed. It seems that ICAT has potent applications to designate appropriate biomarkers for cancer diagnosis.

Stable Isotopic Labeling with Amino Acids in Cell Culture

SILAC is an MS-based approach for quantitative proteomics that depends on metabolic labeling of whole cellular proteome. The proteomes of different cells grown in cell culture are labeled with "light" or "heavy" form of amino acids and differentiated through MS. The SILAC has been developed as an expedient technique to study the regulation of gene expression, cell signaling, post-translational modifications. Additionally, SILAC is a vital technique for secreted pathways and secreted proteins in cell culture.

SILAC was used for quantitative proteome analysis of B. subtilis in two physiological states such as growth during phosphate and succinate starvation. More than 1,500 proteins were identified and quantified in the two tested states. About 75% genes of B. subtilis were expressed in log phase. Moreover, 10 phosphorylation sties were quantified under phosphate starvation while 35 phosphorylation sites under growth on succinate. Highly purified mutant adenovirus deficient in protein V (internal protein component), wild-type adenovirus and recombinant virus were quantified through SILAC. Viral protein composition and abundance were constant in all types of viruses except virus deficient in protein V which also resulted in reduced amount of another viral core protein.

SILAC was used by for quantitative proteome analysis of A. thaliana. Expression of glutathione S-transferase was analyzed in response to abiotic stress due to salicylic acid and consequent proteins were quantified. Salt stress response and protein dynamics in photosynthetic organism Chlamydomonas reinhardtii have been studied to establish the proteome turnover rate and changes in metabolism under salt stress conditions. RuBisCO was found as the most prominent protein in C. reinhardtii.

The intracellular stability of almost 600 proteins from human adenocarcinoma cells have been analyzed through "dynamic SILAC" and the overall protein turnover rate was determined. Tissue regeneration is imperative in many diseases such as lung disease, heart failure and neurodegenerative disorders. The tissue regeneration and protein turnover rate were quantitatively analyzed in zebra fish. Proteome analysis showed that fin, intestine and liver have high regenerative capacity while heart and brain have the lowest. The proteins in tissue regeneration were mainly involved in transport activity and catalytic pathways.

Isobaric Tag for Relative and Absolute Quantitation

iTRAQ is multiplex protein labeling technique for protein quantification based on tandem mass spectrometry. This technique relies on labeling the protein with isobaric tags (8-plex and 4-plex) for relative and absolute quantitation. The technique comprises labeling of the N-terminus and side chain amine groups of proteins, fractionated through liquid chromatography and finally analyzed through MS. It is essential to find the gene regulation to understand the disease mechanism, therefore protein quantitation using iTRAQ is an appropriate method that helps to identify and quantify the protein simultaneously.

iTRAQ has been applied for quantitative analysis of membrane and cellular proteins of Thermobifida fusca grown in the absence and presence of cellulose. About 181 membrane and 783 cytosolic proteins were quantified during cellulosic hydrolysis. The quantified protein in cellulosic medium was involved in pentose phosphate pathway, glycolysis, citric acid cycle, starch, amino acid, fatty acid, purine, pyrimidine and energy metabolism. Consequently, these proteins have a functional role in cell wall synthesis, transcription, translation and replication. The huge amount of oxidative and hydrolytic enzymes is secreted by Phanerochaete chrysosporium that degrade lignin, cellulose and mixture of lignin and cellulose. The secretory proteins were quantified from P. chrysosporium and 117 enzymes were quantified including cellulose hydrolyzing exoglucanases, endoglucanases, cellobiose dehydrogenase and β-glucosidases.

The presence of soluble aluminum ions (Al^{3+}) in soil limits crop growth; however, Oryza sativa are highly aluminum tolerant; therefore, quantitative proteome analysis was carried out in response to Al^{3+} in roots of O. sativa at early stages. Out of 700 identified proteins, the expression of 106 proteins was different in Al^{3+} tolerant and sensitive cultivars. The role of hydrogen peroxide (H_2O_2) in growth of wheat was identified through iTRAQ-based quantitative approach that showed that the increased concentration of H_2O_2 restrained the growth of roots and seedlings of wheat. Out of 3,425 identified proteins, 44 were newly identified H_2O_2- responsive proteins involved in detoxification/stress, carbohydrate metabolism and single transduction. Several proteins such as superoxide dismutase, intrinsic protein 1 and fasciclin-like arabinogalactan protein could possibly be involved in H_2O_2 tolerance.

iTRAQ was a useful tool for determination of molecular process involved in development and function of natural killer (NK) cells. Membrane bound proteins of NK cells from CD3-depleted adult peripheral blood cells and umbilical cord blood stem cells were quantified. Ontology analysis exhibited that many of these proteins were involved in nucleic acid binding, cell signaling and mitochondrial functions. Protein profiling was carried out in mouse liver regeneration following a partial hepatectomy. A total of 827 identified proteins, 270 were quantified as well. Fabp5, Lactb2 and Adh1 were downregulated among these while Pabpc1, Mat1a, Oat, Hpx and Dnpep were upregulated.

X-ray Crystallography

X-ray crystallography is the most preferred technique for three-dimensional structure determination of proteins. The highly purified crystallized samples are exposed to X-rays and the subsequent diffraction patterns are processed to produce information about the size of the repeating unit that forms the crystal and crystal packing symmetry. X-ray crystallography has an extensive range of applications to study the virus system, protein–nucleic acid complexes and immune complexes. Further, the three-dimensional protein structure provides detailed information about the elucidation of enzyme mechanism, drug designing, site-directed mutagenesis and protein–ligand interaction.

ZipA and FtsZ are the vital components of spatial ring structure that facilitates cell division in E. coli. ZipA is a membrane anchored protein while FtsZ is homologous of eukaryotic tubulin and their interaction is facilitated by C-terminal domains. X-ray crystallography revealed the structure of C-terminal fragment of FtsZ and binding complex of FtsZ-ZipA. The structure of Norwalk virus that causes gastroenteritis in humans was determined through X-ray crystallography, which revealed that viral capsid consists of 180 repeating units of single protein. The two domains; shell (S) domain and protruding (P) domain of capsid protein are connected by flexible hinge. Eight-standard β-sandwich motif was present in Shell (S) domain while structure of Protruding (P) domain was similar to the domain of eukaryotic translation elongation factor. These domains are the key determinants responsible of cell binding and strain specificity.

The movement of phospholipids, glycolipids, steroids and fatty acids between membranes occurs due to non-specific lipid transfer proteins (nsLTPs). The comparative structure of maize nsLTP in complex with numerous ligands revealed variations in the volume of the hydrophobic cavity depending on the size of bound ligands. The microsomal cytochrome P450 3A4 catalyzes the drug–drug interaction in humans that induce or inhibit the enzymes and metabolically clear the clinically used drugs. The protein structure was analyzed through X-ray crystallography that exhibited a large substrate binding cavity capable to oxidize huge substrates such as statins, cyclosporin, macrolide antibiotics and taxanes. The X-ray crystallography revealed the 3D structure of recombinant horseradish peroxidase in complex with benzohydroxamic acid (BHA). The electron density for BHA was detected in active site of peroxidase along with hydrophobic pocket adjacent to aromatic ring of the BHA.

High-throughput Techniques

Mass Spectrometry

MS is used to measure the mass to charge ratio (m/z), therefore helpful to determine the molecular weight of proteins. The overall process comprises three steps. The molecules must be transformed to gas-phase ions in the first step, which poses a challenge for biomolecules in a liquid or solid phase. The second step involves the separation of ions on the basis of m/z values in the presence of electric or magnetic fields in a compartment known as mass analyzer. Finally, the separated ions and the amount of each species with a particular m/z value are measured. Commonly used ionization method comprises matrix-assisted laser desorption ionization (MALDI), surface enhanced laser desorption/Ionization (SELDI) and electrospray ionization (ESI).

In clinical laboratories, bacterial identification depends on conventional techniques. However, identification of slow growing, fastidious and anaerobic bacteria through conventional techniques is expensive, complex and time consuming. Biswas and Rolain used the MALDI-TOF for early pathogenic bacterial identification, which is useful for early disease control. MS has also became an significant tool in virus research at molecular level, and various viruses and viral proteins including intact viruses, mutant viral strains, capsid protein, post-translational modifications were identified. The study of the changes of viral capsid protein during the infection has allowed the researcher to develop new antiviral drugs. Electrospray ionization mass spectrometry (ESI-MS) coupled with PCR and rRNA gene sequencing provided the accurate and rapid identification of medically important filamentous fungi, yeast and Prototheca species.

Post-translational modification in plants including protein phosphorylation has been distinguished through MS. Top down Fourier Transform mass spectrometry was used to the characterize chloroplast proteins of A. thaliana. Hydrophobic properties and molecular mass of light harvesting proteins of photosystem-II of 14 different plants species were presented by Zolla et al ESI-MS was used for profiling of integral membrane proteins and detection of post-translational modifications. The most abundant proteins of tomato (Lycopersicon esculentum) xylem sap after Fusarium oxysporum infection were detected with mass spectrometric sequencing and peptide mass finger printing.

The blood proteins including the IBP2, IBP3, IGF1, IGF2 and A2GL have been proposed as biomarkers for the diagnosis of breast cancer. MS was used to characterize these blood proteins. PSA, human growth hormone and interleukin-12 were also analyzed from human serum. Imaging MALDI mass spectrometry was used for the analysis of whole body tissues. The distribution of drugs and metabolites was detected within whole body tissues following drug administration that was useful to analyze novel therapeutics and provide deeper insight into toxicological and therapeutic process.

NMR Spectroscopy

The NMR is a leading tool for the investigation of molecular structure, folding and behavior of proteins. Structure determination through NMR spectroscopy typically involves various phases, each using a discrete set of extremely specific techniques. The samples are prepared and measurements are made followed by interpretive approaches to confirm the structure. The protein structure is fundamental in several research areas such as structure-based drug design, homology modeling and functional genomics.

The three-dimensional structure of transmembrane domain of outer membrane protein A from E. coli has been determined through heteronuclear NMR in dodecylphosphocholine micelles. The fold of protein consists of 19 kDa (177 amino acids) and the structure comprises larger mobile loops toward extracellular side and an eight-stranded β-barrel linked by tight turns on the periplasmic side. The interaction of iso-1-cytochrom c with cytochrome c peroxidase from yeast was investigated by NMR. Chemical shift was observed for both ^{1}H and ^{15}N nuclei arising from the interface of isotopically enriched ^{15}N cytochrome c with cytochrome c peroxidase.

Plant litter decomposition is essential in nitrogen and carbon cycles for the provision of necessary nutrients to the soil and atmospheric CO_2. ^{15}N- and ^{13}C-labeled plant materials were used to monitor the environmental degradation of wheatgrass and pine residues via HR-MAS NMR spectroscopy. The spectra revealed that condensed and hydrolysable tannin were lost from all plant tissues whereas the aliphatic components (cuticles, waxes) and aromatic (partly lignin) persisted along with a small portion of carbohydrate.

Holmes et al. described the variations between metabolic phenotypic from 4,630 participants belonging to 4 human populations through NMR spectroscopy. Metabolic phenotypes including in the study were the products of interactions between variety of factors such as environmental, dietary, genetic and gut microbial activities. Selective metabolites across populations were associated with blood pressure and urinary metabolites that offer the promising discovery of novel biomarkers.

The NMR can be coupled with various approaches like LC or UHPLC to increase the resolution and sensitivity for high-throughput protein profiling. In addition, the structural information can be generated is compared in relation to the identification of metabolites in complex mixtures. NMR coupled with ultra-high performance liquid chromatography (UHPLC) was developed to characterize the metabolic disturbances in esophageal cancer patients for the identification of possible biomarkers for early diagnosis and prognosis. The study revealed considerable alterations in ketogenesis, glycolysis and tricarboxylic acid cycle and amino acid and lipid metabolism in esophageal cancer patients compared with the controls.

Bioinformatics Analysis

Bioinformatics is an essential component of proteomics; therefore, its implications have been progressively increasing with the advent of high-throughput methods that are dependent on powerful data analysis. This new and emergent field is presenting novel algorithms to manage huge and heterogeneous proteomics data and headway toward the discovery procedure.

Endolysins are class of antibacterial enzymes that are becoming useful tool to control spreading of multi-resistant bacteria. The antibacterial property can be altered or expanded by domain swapping, mutagenesis or gene shuffling. The challenge of designing specific endolysins has been revealed in-silico analysis for protein domains present in prophage and phage endolysins. The combination of domains have been studied and sequence type with domain arrangement and conserved amino acids have been determined through multiple sequence alignment. The presence, number and types of binding domain with in endolysins sequence also have been studied. In-silico analysis approach was used to calculate the distribution of the plant food allergens into protein families and determination of conserved surface essential for IgE cross reactivity. The plant food allergen sequences were categorized into four families that indicate the role of conserved structures and biological activities in stimulating allergic properties.

A blood coagulation enzyme, Human Factor Xa (FXa) catalyzes the activation of prothrombin to thrombin and plays an important role in thrombosis and hemostasis. The imbalance in the activation of enzymes intrudes the hemostasis leading to the blood disorders. The safe and effective anticoagulants may be developed by direct inhibition of FXa without effecting thrombin activity essential for normal hemostasis. A study aided the design of more effective ligands through Discovery Studio. Docking studies and binding confirmations revealed that sulfonamide derivatives were inhibitors of FXa.

The use of Bioinformatics for proteomics has gain significantly affluent during the previous few years. The development of new algorithm for the analysis of higher amount of data with increased specificity and accuracy helps in the identification and quantitation of proteins therefore have made possible to achieve expounded data regarding protein expression. The management of such a high quantity of data is the main problem associated with these kind of analyses. Further, it is still difficult to find the association between proteomic data and the other omics technologies including genomics and metabolomics. The database technology along with new semantic statistical algorithms however are the potent tools that might be useful to overcome these limitations.

For MS, the proteins are extracted from the sample and digested using one or several proteases to produce definite set of peptides. Further steps including enrichment and fractionation can be added at protein or peptide level to decrease the complexity of sample or when the analysis of specific subset of proteins is desired. The obtained peptides are analyzed by liquid chromatography coupled with mass spectrometry (LC–MS). Common approaches include either the analysis of deep coverage of proteome by shotgun MS or quantitative investigation for a definite set of proteins through targeted MS. The resulting spectra provide information regarding the sequence, which is important for the identification of proteins. The obtained data may be displayed in a form of 3-D map with mass-to-charge (m/z) ratio, retention time (RT) and intensities of peptides along with fragmentation spectra. The intensity of mass to charge ratio for a particular peptide is plotted along the RTs to get the chromatographic peak. The area under this curve can be used for quantification of peptides, whereas the proteins are identified by the fragmentation spectra. The proteomic data can be uploaded to the repositories that can also be helpful for searching the database. The largest proteome repositories including PRIDE proteomics identification database, Proteome Commons and PeptideAtlas project provide direct access to most of stored data and are valuable tools for data mining.

The protein pathways are a series of reactions inside the cell that exert a particular biological effect. The proteins that are directly involved in reaction along with those that regulates the pathways are combined in pathway databases; therefore, a number of resources and databases are available for the protein pathways. The KEGG, Ingenuity, Pathway Knowledge Base Reactome and BioCarta are some of the pathway databases that include a comprehensive data regarding metabolism, signaling and interactions. In addition to these comprehensive databases, the specific databases for signal transduction pathways such as GenMAPP or PANTHER have been developed. Moreover, databases such as Netpath have been developed, which involve the pathways active in cancer that are helpful for the identification of proteins relevant for a cancer type. These public databases possess higher connectivity that allows novel findings for proteins.

The proteins do not act independently in most of the cases and form transient or stable complexes with other proteins. The protein might be intricate as complexes of variable composition and it is essential to study the protein complexes along with the conditions that result in their formation or dissociation for the complete understanding of a biological system. The databases such as BioGRID, IntAct, MINT and HRPD contain the information with reference to protein interactions in complexes. STRING is not only a widely used database for protein interaction data, but it connects to various other resources for literature mining. Furthermore, protein networks can be drawn based on the list of genes provided and the available interactions using STRING database.

Sample Preparation for Proteomics

Preparation of sample is the most fundamental step in proteomics research that considerably affects the results of an experiment. Therefore, the selection of appropriate experimental model and sample preparation method is essential for reliable results, especially in comparative proteomics, that deal with the minor variances of experimental samples compared with the control. The major impediments associated with the analysis of complex biological materials are the wider range of protein abundance. A particular cell could have only few copies of a protein, but we may expect up to million copies of an abundant protein therefore these abundant proteins should be removed for most of the proteomic analysis. The Pre analytical samples treatment include various methods for fractionation and proteins enrichment could be helpful in this regard.

The animal tissue associated with the disease is often selected for proteomic analysis after the establishment of particular animal model. The tissue characteristics vary among the types, for example brain tissue have abundance of lipids that need to be eliminated for high quality results. The selective precipitation of proteins with acetone and trichloroacetic acid (TCA) is a widely used method for protein expression profiling in neuroscience. The fresh tissue samples are usually perfused with cold saline before excision and are used as unfettered from fat as well as connective tissue. The tissue is minced in freshly prepared lysis buffer that might contain detergents and/or protease inhibitors. The biopsy is frequently used as a source of tissue for expression analysis that is usually surgically obtained and need to freeze immediately using liquid nitrogen and stored at −80 °C before analysis.

The plant cells have the distinctive cell wall made up of cellulose mainly and its derivatives. The primary cell wall surrounds the young plant cells although some type of plants and cells contains a rigid secondary cell wall after developmental phase. The release of proteins as a result of cell wall disruption is essential for analytical success; therefore, different physical and chemical techniques are employed for the destruction of cell wall, for example, freeze thawing, sonication, high speed blending and use of lysing buffer. The CWPs constitute ~10% of the cell wall mass and are mainly involved in signaling, modification of cell wall constituents and communications with plasma membrane proteins. The extraction of CWPs is challenging and the available cell wall proteomes so far contain either labile or loosely bound proteins.

The majority of research is conducted on model plants, i.e., rice (O. sativa) and Thale cress (A. thaliana) having a relatively small genome. Another problem associated with plants proteome analysis is the presence of contaminants other than proteins specific to the plant type including lipids, organic acids, polyphenols, terpenes and pigments that can impede in the separation procedures. The cleaning procedures are therefore desirable that frequently uses acetone and TCA. It is established that TCA alone is insufficient to remove contaminants and therefore sonication and brief grinding are suggested along with TCA.

The variable pI range of proteins, their relative abundance, hydrophobicity and solubility makes them difficult to separate through the classical 2-DE. The liquid chromatography technique connected with MS (LC–MS/MS) can be used as an alternative separation method. The sample preparation procedure in plant proteomics is generally dependent on the type of plants, its fragment (leaf, stem, fruit, etc.) and even on the stage of plant development. Fukuda et al. described the protocol for the preparation of sample from rice embryo and its analysis using 2D electrophoresis. The plant material was chemically homogenized with solution consisting of urea, thiourea, CHAPS (3-[(3-Cholamidopropyl)-dimethyl-ammonio] 1-propane sulfonate), Ampholine, polyvinyl lopolypyrrolidone and 2-mercaptoethanol. The mixture was boiled at 100 °C, centrifuged and supernatant was discarded. Finally, the lipids were removed with the addition of hexane, and the samples were analyzed by 2D electrophoresis.

Protein: Types and Processes

The large biomolecules or macromolecules composed of one or more long chains of amino acid which are connected through a covalent peptide bond are known as proteins. They are classified into different types such as blood proteins, globular protein, GST-tagged proteins and HIS-tagged proteins. These diverse classifications of proteins have been thoroughly discussed in this chapter.

Proteins are long chains of amino acids that form the basis of all life. They are like machines that make all living things, whether viruses, bacteria, butterflies, jellyfish, plants, or human function.

The human body consists of around 100 trillion cells. Each cell has thousands of different proteins. Together, these cause each cell to do its job. The proteins are like tiny machines inside the cell.

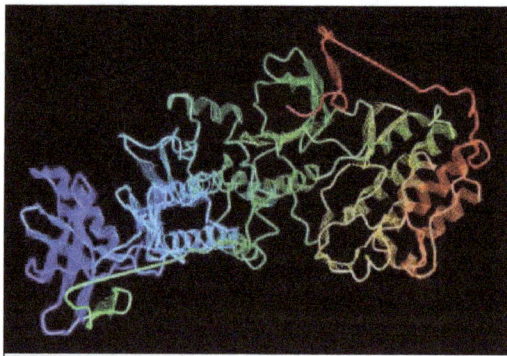

Protein molecules are essential for the functioning of every
cell in the body. The body synthesizes some proteins foods we eat.

Amino Acids and Proteins

Protein consists of amino acids, and amino acids are the building blocks of protein. There are around 20 amino acids.

These 20 amino acids can be arranged in millions of different ways to create millions of different proteins, each with a specific function in the body. The structures differ according to the sequence in which the amino acids combine.

The 20 different amino acids that the body uses to synthesize proteins are: Alanine, arginine, asparagine, aspartic acid, cysteine, glutamic acid, glutamine, glycine, histidine, isoleucine, leucine, lysine, methionine, phenylalanine, proline, serine, threonine, tryptophan, tyrosine, and valine.

Amino acids are organic molecules that consist of carbon, hydrogen, oxygen, nitrogen, and sometimes sulfur.

It is the amino acids that synthesize proteins and other important compounds in the human body, such as creatine, peptide hormones, and some neurotransmitters.

Types of Protein

We sometimes hear that there are three types of protein foods:

- Complete proteins: These foods contain all the essential amino acids. They mostly occur in animal foods, such as meat, dairy, and eggs.

- Incomplete proteins: These foods contain at least one essential amino acid, so there is a lack of balance in the proteins. Plant foods, such as peas, beans, and grains mostly contain incomplete protein.

- Complementary proteins: These refer to two or more foods containing incomplete proteins that people can combine to supply complete protein. Examples include rice and beans or bread with peanut butter.

What do Proteins do?

Proteins play a role in nearly every biological process, and their functions vary widely.

The main functions of proteins in the body are to build, strengthen and repair or replace things, such as tissue.

They can be:

- Structural, like collagen.
- Hormonal, like insulin.
- Carriers, for example, hemoglobin.
- Enzymes, such as amylase.

All of these are proteins.

Keratin is a structural protein that strengthens protective coverings, such as hair. Collagen and elastin, too, have a structural function, and they also provide support for connective tissue.

Most enzymes are proteins and are catalysts, which means they speed up chemical reactions. They are necessary for respiration in human cells, for example, or photosynthesis in plants.

Rice and beans together provide complete protein.

Protein is one of the essential nutrients, or macronutrients, in the human diet, but not all the protein we eat converts into proteins in our body.

When people eat foods that contain amino acids, these amino acids make it possible for the body to create, or synthesize, proteins. If we do not consume some amino acids, we will not synthesize enough proteins for our bodies to function correctly.

There are also nine essential amino acids that the human body does not synthesize, so they must come from the diet. All food proteins contain some of each amino acid, but in different proportions.

Gelatin is special in that it contains a high proportion of some amino acids but not the whole range. The nine essential acids that the human body does not synthesize are: histidine, isoleucine, leucine, lysine, methionine, phenylalanine, threonine, tryptophan, and valine.

Foods that contain these nine essential acids in roughly equal proportions are called complete proteins. Complete proteins mainly come from animal sources, such as milk, meat, and eggs. Soy and quinoa are vegetable sources of complete protein. Combining red beans or lentils with wholegrain rice or peanut butter with wholemeal bread also provides complete protein.

The body does not need all the essential amino acids at each meal, because it can utilize amino acids from recent meals to form complete proteins. If you have enough protein throughout the day, there is no risk of a deficiency. In other words, the recommended nutrient is protein, but what we really need is amino acids.

INTRINSICALLY DISORDERED PROTEINS

An intrinsically disordered protein (IDP) is a protein that lacks a fixed or ordered three-dimensional structure. IDPs cover a spectrum of states from fully unstructured to partially structured and include random coils, (pre-)molten globules, and large multi-domain proteins connected by flexible linkers. They constitute one of the main types of protein (alongside globular, fibrous and membrane proteins).

The discovery of IDPs has challenged the traditional protein structure paradigm, that protein function depends on a fixed three-dimensional structure. This dogma has been challenged over the 2000s and 2010s by increasing evidence from various branches of structural biology, suggesting that protein dynamics may be highly relevant for such systems. Despite their lack of stable structure, IDPs are a very large and functionally important class of proteins. In some cases, IDPs can adopt a fixed three-dimensional structure after binding to other macromolecules. Overall, IDPs are different from structured proteins in many ways and tend to have distinct properties in terms of function, structure, sequence, interactions, evolution and regulation.

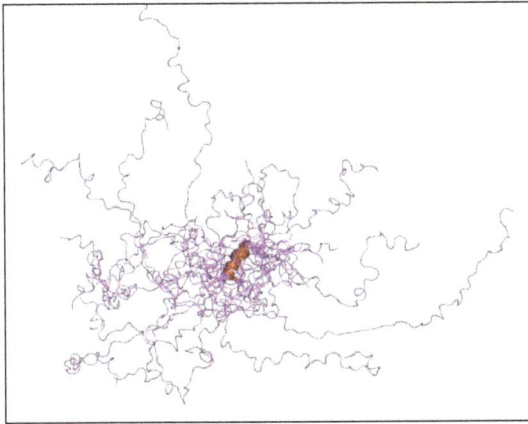

An ensemble of NMR structures of the Thylakoid soluble phosphoprotein TSP9, which shows a largely flexible protein chain.

In the 1930s -1950s, the first protein structures were solved by protein crystallography. These early structures suggested that a fixed three-dimensional structure might be generally required to mediate biological functions of proteins. When stating that proteins have just one uniquely defined configuration, Mirsky and Pauling did not recognize that Fisher's work would have supported their thesis with his 'Lock and Key' model. These publications solidified the central dogma of molecular biology in that the sequence determines the structure which, in turn, determines the function of proteins. In 1950, Karush wrote about 'Configurational Adaptability' contradicting all the assumptions and research in the 19th century. He was convinced that proteins have more than one configuration at the same energy level and can choose one when binding to other substrates. In the 1960s, Levinthal's paradox suggested that the systematic conformational search of a long polypeptide is unlikely to yield a single folded protein structure on biologically relevant timescales (i.e. seconds to minutes). Curiously, for many (small) proteins or protein domains, relatively rapid and efficient refolding can be observed in vitro. As stated in Anfinsen's Dogma from 1973, the fixed 3D structure of these proteins is uniquely encoded in its primary structure (the amino acid sequence), is kinetically accessible and stable under a range of (near) physiological conditions, and can therefore be considered as the native state of such "ordered" proteins.

During the subsequent decades, however, many large protein regions could not be assigned in x-ray datasets, indicating that they occupy multiple positions, which average out in electron density maps. The lack of fixed, unique positions relative to the crystal lattice suggested that these regions were "disordered". Nuclear magnetic resonance spectroscopy of proteins also demonstrated the presence of large flexible linkers and termini in many solved structural ensembles. It is now generally accepted that proteins exist as an ensemble of similar structures with some regions more constrained than others. Intrinsically Unstructured Proteins (IUPs) occupy the extreme end of this spectrum of flexibility, whereas IDPs also include proteins of considerable local structure tendency or flexible multidomain assemblies. These highly dynamic disordered regions of proteins have subsequently been linked to functionally important phenomena such as allosteric regulation and enzyme catalysis.

In the 2000s, bioinformatic predictions of intrinsic disorder in proteins indicated that intrinsic disorder is more common in sequenced/predicted proteomes than in known structures in the protein database. Based on DISOPRED2 prediction, long (>30 residue) disordered segments occur in 2.0% of archaean, 4.2% of eubacterial and 33.0% of eukaryotic proteins. In 2001, Dunker published his paper 'Intrinsically Disordered Proteins' questioning whether the newly found information was ignored for 50 years.

In the 2010s it became clear that IDPs are highly abundant among disease-related proteins.

Biological Roles

Many disordered proteins have the binding affinity with their receptors regulated by post-translational modification, thus it has been proposed that the flexibility of disordered proteins facilitates the different conformational requirements for binding the modifying enzymes as well as their receptors. Intrinsic disorder is particularly enriched in proteins implicated in cell signaling, transcription and chromatin remodeling functions. Genes that have recently been born de novo tend to have higher disorder.

Flexible Linkers

Disordered regions are often found as flexible linkers or loops connecting domains. Linker sequences vary greatly in length but are typically rich in polar uncharged amino acids. Flexible linkers allow the connecting domains to freely twist and rotate to recruit their binding partners via protein domain dynamics. They also allow their binding partners to induce larger scale conformational changes by long-range allostery.

Linear Motifs

Linear motifs are short disordered segments of proteins that mediate functional interactions with other proteins or other biomolecules RNA, DNA, sugars etc.

Many roles of linear motifs are associated with cell regulation, for instance in control of cell shape, subcellular localisation of individual proteins and regulated protein turnover. Often, post-translational modifications such as phosphorylation tune the affinity (not rarely by several orders of magnitude) of individual linear motifs for specific interactions. Relatively rapid evolution and a relatively small number of structural restraints for establishing novel (low-affinity) interfaces make it particularly challenging to detect linear motifs but their widespread biological roles and the fact that many viruses mimick/hijack linear motifs to efficiently recode infected cells underlines the timely urgency of research on this very challenging and exciting topic. Unlike globular proteins IDPs do not have spatially-disposed active pockets. Nevertheless, in 80% of IDPs (~3 dozens) subjected to detailed structural characterization by NMR there are linear motifs termed PreSMos (pre-structured motifs) that are transient secondary structural elements primed for target recognition. In several cases it has been demonstrated that these transient structures become full and stable secondary structures, e.g., helices, upon target binding. Hence, PreSMos are the putative active sites in IDPs.

Coupled Folding and Binding

Many unstructured proteins undergo transitions to more ordered states upon binding to their targets (e.g. Molecular Recognition Features (MoRFs)). The coupled folding and binding may be local, involving only a few interacting residues, or it might involve an entire protein domain. It was recently shown that the coupled folding and binding allows the burial of a large surface area that would be possible only for fully structured proteins if they were much larger. Moreover, certain disordered regions might serve as "molecular switches" in regulating certain biological function by switching to ordered conformation upon molecular recognition like small molecule-binding, DNA/RNA binding, ion interactions etc.

The ability of disordered proteins to bind, and thus to exert a function, shows that stability is not a required condition. Many short functional sites, for example Short Linear Motifs are over-represented in disordered proteins. Disordered proteins and short linear motifs are particularly abundant in many RNA viruses such as Hendra virus, HCV, HIV-1 and human papillomaviruses. This enables such viruses to overcome their informationally limited genomes by facilitating binding, and manipulation of, a large number of host cell proteins.

Disorder in the Bound State

Intrinsically disordered proteins can retain their conformational freedom even when they bind specifically to other proteins. The structural disorder in bound state can be static or dynamic. In fuzzy complexes structural multiplicity is required for function and the manipulation of the bound disordered region changes activity. The conformational ensemble of the complex is modulated via post-translational modifications or protein interactions. Specificity of DNA binding proteins often depends on the length of fuzzy regions, which is varied by alternative splicing.

Structural Aspects

Intrinsically disordered proteins adapt many different structures in vivo according to the cell's conditions, creating a structural or conformational ensemble.

Therefore, their structures are strongly function-related. However, only few proteins are fully disordered in their native state. Disorder is mostly found in intrinsically disordered regions (IDRs) within an otherwise well-structured protein. The term intrinsically disordered protein (IDP) therefore includes proteins that contain IDRs as well as fully disordered proteins.

The existence and kind of protein disorder is encoded in its amino acid sequence. In general, IDPs are characterized by a low content of bulky hydrophobic amino acids and a high proportion of polar and charged amino acids, usually referred to as low hydrophobicity. This property leads to good interactions with water. Furthermore, high net charges promote disorder because of electrostatic repulsion resulting from equally charged residues. Thus disordered sequences cannot sufficiently bury a hydrophobic core to fold into stable globular proteins. In some cases, hydrophobic clusters in disordered sequences provide the clues for identifying the regions that undergo coupled folding and binding. Many disordered proteins reveal regions without any regular secondary structure These regions can be termed as flexible, compared to structured loops. While the latter are rigid and contain only one set of Ramachandran angles, IDPs involve multiple sets of angles. The term flexibility is also used for well-structured proteins, but describes a different phenomenon in the context of disordered proteins. Flexibility in structured proteins is bound to an equilibrium state, while it is not so in IDPs. Many disordered proteins also reveal low complexity sequences, i.e. sequences with over-representation of a few residues. While low complexity sequences are a strong indication of disorder, the reverse is not necessarily true, that is, not all disordered proteins have low complexity sequences. Disordered proteins have a low content of predicted secondary structure.

Experimental Validation

Large-scale in-cell validation of IDR predictions is possible using biotin 'painting'. Intrinsically unfolded proteins, once purified, can be identified by various experimental methods. The primary method to obtain information on disordered regions of a protein is NMR spectroscopy. The lack of electron density in X-ray crystallographic studies may also be a sign of disorder.

Folded proteins have a high density (partial specific volume of 0.72-0.74 mL/g) and commensurately small radius of gyration. Hence, unfolded proteins can be detected by methods that are sensitive to molecular size, density or hydrodynamic drag, such as size exclusion chromatography, analytical ultracentrifugation, small angle X-ray scattering (SAXS), and measurements of the diffusion constant. Unfolded proteins are also characterized by their lack of secondary structure, as assessed by far-UV (170-250 nm) circular dichroism (esp. a pronounced minimum at ~200 nm) or infrared spectroscopy. Unfolded proteins also have exposed backbone peptide groups exposed to solvent, so that they are readily cleaved by proteases, undergo rapid hydrogen-deuterium exchange and exhibit a small dispersion (<1 ppm) in their 1H amide chemical shifts as measured by NMR. (Folded proteins typically show dispersions as large as 5 ppm for the amide protons.) Recently, new methods including Fast parallel proteolysis (FASTpp) have been introduced, which allow to determine the fraction folded/disordered without the need for purification. Even subtle differences in the stability of missense mutations, protein partner binding and (self)polymerisation-induced folding of (e.g.) coiled-coils can be detected using FASTpp as recently demonstrated using the tropomyosin-troponin protein interaction. Fully unstructured protein regions can be experimentally validated by their hypersusceptibility to proteolysis using short digestion times and low protease concentrations.

Bulk methods to study IDP structure and dynamics include SAXS for ensemble shape information, NMR for atomistic ensemble refinement, Fluorescence for visualising molecular interactions and conformational transitions, x-ray crystallography to highlight more mobile regions in otherwise rigid protein crystals, cryo-EM to reveal less fixed parts of proteins, light scattering to monitor size distributions of IDPs or their aggregation kinetics, Circular Dichroism to monitor secondary structure of IDPs.

Single-molecule methods to study IDPs include spFRET to study conformational flexibility of IDPs and the kinetics of structural transitions, optical tweezers for high-resolution insights into the ensembles of IDPs and their oligomers or aggregates, nanopores to reveal global shape distributions of IDPs, magnetic tweezers to study structural transitions for long times at low forces, high-speed AFM to visualise the spatio-temporal flexibility of IDPs directly.

Disorder Annotation

Intrinsic disorder can be either annotated from experimental information or predicted with specialized software. Disorder prediction algorithms can predict Intrinsic Disorder (ID) propensity with high accuracy (approaching around 80%) based on primary sequence composition, similarity to unassigned segments in protein x-ray datasets, flexible regions in NMR studies and physico-chemical properties of amino acids.

Compilation of screenshots from PDB database and molecule representation via VMD. Blue and red arrows point to missing residues on receptor and growth hormone, respectively.

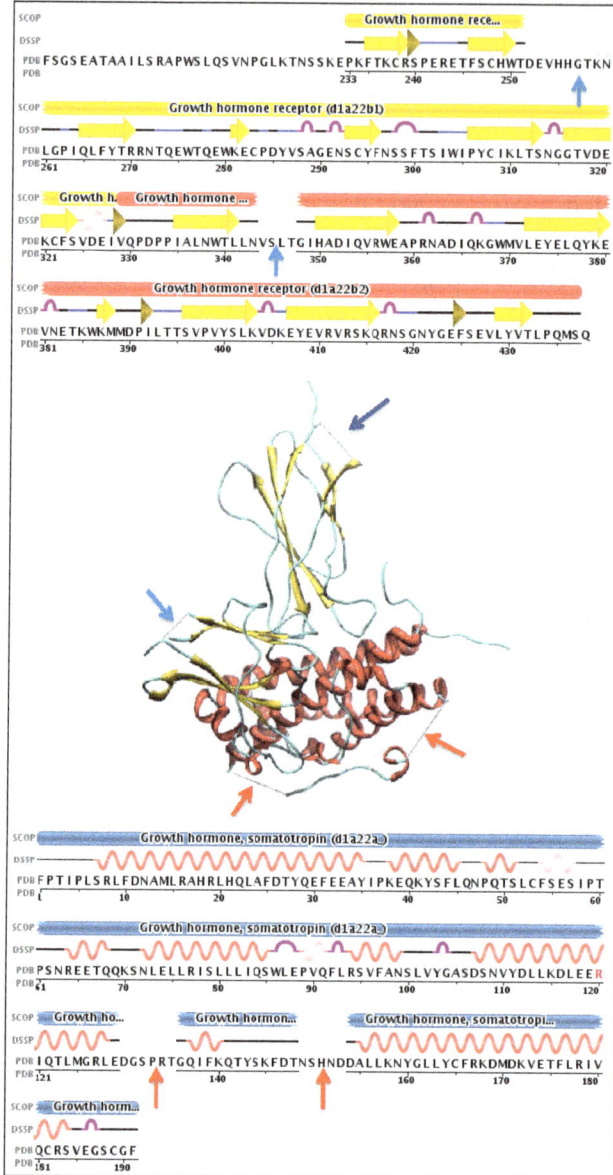

Missing electron densities in X-ray
structure representing protein disorder (PDB: 1a22,
human growth hormone bound to receptor).

Disorder Databases

Databases have been established to annotate protein sequences with intrinsic disorder information. The DisProt database contains a collection of manually curated protein segments which have been experimentally determined to be disordered. MobiDB is a database combining exoerimentally curated disorder annotations (e.g. from DisProt) with data derived from missing residues in X-ray crystallographic structures and flexible regions in NMR structures.

Distinguishing IDPs from Well-structured Proteins

Separating disordered from ordered proteins is essential for disorder prediction. One of the first steps to find a factor that distinguishes IDPs from non-IDPs is to specify biases within the amino acid composition. The following hydrophilic, charged amino acids A, R, G, Q, S, P, E and K have been characterized as disorder-promoting amino acids, while order-promoting amino acids W, C, F, I, Y, V, L, and N are hydrophobic and uncharged. The remaining amino acids H, M, T and D are ambiguous, found in both ordered and unstructured regions. This information is the basis of most sequence-based predictors. Regions with little to no secondary structure, also known as NORS (NO Regular Secondary structure) regions, and low-complexity regions can easily be detected. However, not all disordered proteins contain such low complexity sequences.

Prediction Methods

Determining disordered regions from biochemical methods is very costly and time-consuming. Due to the variable nature of IDPs, only certain aspects of their structure can be detected, so that a full characterization requires a large number of different methods and experiments. This further increases the expense of IDP determination. In order to overcome this obstacle, computer-based methods are created for predicting protein structure and function. It is one of the main goals of bioinformatics to derive knowledge by prediction. Predictors for IDP function are also being developed, but mainly use structural information such as linear motif sites. There are different approaches for predicting IDP structure, such as neural networks or matrix calculations, based on different structural and biophysical properties.

Many computational methods exploit sequence information to predict whether a protein is disordered. Notable examples of such software include IUPRED and Disopred. Different methods may use different definitions of disorder. Meta-predictors show a new concept, combining different primary predictors to create a more competent and exact predictor.

Due to the different approaches of predicting disordered proteins, estimating their relative accuracy is fairly difficult. For example, neural networks are often trained on different datasets. The disorder prediction category is a part of biannual CASP experiment that is designed to test methods according accuracy in finding regions with missing 3D structure (marked in PDB files as REMARK465, missing electron densities in X-ray structures).

Disorder and Disease

Intrinsically unstructured proteins have been implicated in a number of diseases. Aggregation of misfolded proteins is the cause of many synucleinopathies and toxicity as those proteins start binding to each other randomly and can lead to cancer or cardiovascular diseases. Thereby, misfolding can happen spontaneously because millions of copies of proteins are made during the lifetime of an organism. The aggregation of the intrinsically unstructured protein α-synuclein is thought to be responsible. The structural flexibility of this protein together with its susceptibility to modification in the cell leads to misfolding and aggregation. Genetics, oxidative and nitrative stress as well as mitochondrial impairment impact the structural flexibility of the unstructured α-synuclein protein and associated disease mechanisms. Many key tumour suppressors have large intrinsically unstructured regions, for example p53 and BRCA1. These regions of the proteins are responsible for mediating many of their interactions. Taking the cell's native defense mechanisms as a model drugs can be developed, trying to block the place of noxious substrates and inhibiting them, and thus counteracting the disease.

Computer Simulations

Owing to high structural heterogeneity, NMR/SAXS experimental parameters obtained will be an average over a large number of highly diverse and disordered states (an ensemble of disordered states). Hence, to understand the structural implications of these experimental parameters, there is a necessity for accurate representation of these ensembles by computer simulations. All-atom molecular dynamic simulations can be used for this purpose but their use is limited by the accuracy of current force-fields in representing disordered proteins. Nevertheless, some force-fields have been explicitly developed for studying disordered proteins by optimising force-field parameters using available NMR data for disordered proteins.

MD simulations restrained by experimental parameters (restrained-MD) have also been used to characterise disordered proteins. In principle, one can sample the whole conformational space given an MD simulation (with accurate Force-field) is run long enough. Because of very high structural heterogeneity, the time scales that needs to be run for this purpose are very large and are limited by computational power. However, other computational techniques such as accelerated-MD simulations, replica exchange simulations, metadynamics, multicanonical MD simulations, or methods using coarse-grained representation have been used to sample broader conformational space in smaller time scales.

Moreover, various protocols and methods of analyzing IDPs, such as studies based on quantitative analysis of GC content in genes and their respective chromosomal bands, have been used to understand functional IDP segments.

BLOOD PROTEINS

Blood proteins, also termed plasma proteins, are proteins present in blood plasma. They serve many different functions, including transport of lipids, hormones, vitamins and minerals in activity and functioning of the immune system. Other blood proteins act as enzymes, complement components, protease inhibitors or kinin precursors.

Contrary to popular belief, haemoglobin is not a blood protein, as it is carried within red blood cells, rather than in the blood serum.

Serum albumin accounts for 55% of blood proteins, is a major contributor to maintaining the oncotic pressure of plasma and assists, as a carrier, in the transport of lipids and steroid hormones. Globulins make up 38% of blood proteins and transport ions, hormones, and lipids assisting in immune function. Fibrinogen comprises 7% of blood proteins; conversion of fibrinogen to insoluble fibrin is essential for blood clotting. The remainder of the plasma proteins (1%) are regulatory proteins, such as enzymes, proenzymes, and hormones. All blood proteins are synthesized in liver except for the gamma globulins.

Separating serum proteins by electrophoresis is a valuable diagnostic tool as well as a way to monitor clinical progress. Current research regarding blood plasma proteins is centered on performing proteomics analyses of serum/plasma in the search for biomarkers. These efforts started with two-dimensional gel electrophoresis efforts in the 1970s and in more recent times this research has been performed using LC-tandem MS based proteomics. The normal laboratory value of serum total protein is around 7 g/dL.

Families of Blood Proteins

Blood protein	Normal level	%	Function
Albumins	3.5-5.0 g/dl	55%	create and maintain oncotic pressure; transport insoluble molecules
Globulins	2.0-2.5 g/dl	38%	participate in immune system
Fibrinogen	0.2-0.45 g/dl	7%	Blood coagulation
Regulatory proteins		<1%	Regulation of gene expression
Clotting factors		<1%	Conversion of fibrinogen into fibrin

Examples of specific blood proteins:

- Prealbumin (transthyretin),

- Alpha 1 antitrypsin (neutralizes trypsin that has leaked from the digestive system),

- Alpha 1 acid glycoprotein,

- Alpha 1 fetoprotein,

- alpha2-macroglobulin,

- Gamma globulins,

- Beta-2 microglobulin,

- Haptoglobin,

- Ceruloplasmin,

- Complement component 3,

- Complement component 4

- C-reactive protein (CRP),

- Lipoproteins (chylomicrons, VLDL, LDL, HDL),

- Transferrin,

- Prothrombin,

- MBL or MBP.

GLOBULAR PROTEIN

Globular proteins or spheroproteins are spherical ("globe-like") proteins and are one of the common protein types (the others being fibrous, disordered and membrane proteins). Globular proteins are somewhat water-soluble (forming colloids in water), unlike the fibrous or membrane proteins. There are multiple fold classes of globular proteins, since there are many different architectures that can fold into a roughly spherical shape. The term globin can refer more specifically to proteins including the globin fold.

Globular Structure and Solubility

The term globular protein is quite old (dating probably from the 19th century) and is now somewhat archaic given the hundreds of thousands of proteins and more elegant and descriptive structural motif vocabulary. The globular nature of these proteins can be determined without the means of modern techniques, but only by using ultracentrifuges or dynamic light scattering techniques.

The spherical structure is induced by the protein's tertiary structure. The molecule's apolar (hydrophobic) amino acids are bounded towards the molecule's interior whereas polar (hydrophilic) amino acids are bound outwards, allowing dipole-dipole interactions with the solvent, which explains the molecule's solubility.

Globular proteins are only marginally stable because the free energy released when the protein folded into its native conformation is relatively small. This is because protein folding requires entropic cost. As a primary sequence of a polypeptide chain can form numerous conformations, native globular structure restricts its conformation to a few only. It results in a decrease in randomness, although non-covalent interactions such as hydrophobic interactions stabilize the structure.

Although it is still unknown how proteins fold up naturally, new evidence has helped advance understanding. Part of the protein folding problem is that several non-covalent, weak interactions are formed, such as hydrogen bonds and Van der Waals interactions. Via several techniques, the mechanism of protein folding is currently being studied. Even in the protein's denatured state, it can be folded into the correct structure.

Globular proteins seem to have two mechanisms for protein folding, either the diffusion-collision model or nucleation condensation model, although recent findings have shown globular proteins, such as PTP-BL PDZ2, that fold with characteristic features of both models. These new findings have shown that the transition states of proteins may affect the way they fold. The folding of globular proteins has also recently been connected to treatment of diseases, and anti-cancer ligands have been developed which bind to the folded but not the natural protein. These studies have shown that the folding of globular proteins affects its function.

By the second law of thermodynamics, the free energy difference between unfolded and folded states is contributed by enthalpy and entropy changes. As the free energy difference in a globular protein that results from folding into its native conformation is small, it is marginally stable, thus providing a rapid turnover rate and effective control of protein degradation and synthesis.

Role

Unlike fibrous proteins which only play a structural function, globular proteins can act as:

- Enzymes, by catalyzing organic reactions taking place in the organism in mild conditions and with a great specificity. Different esterases fulfill this role.

- Messengers, by transmitting messages to regulate biological processes. This function is done by hormones, i.e. insulin etc.

- Transporters of other molecules through membranes.

- Stocks of amino acids.

- Regulatory roles are also performed by globular proteins rather than fibrous proteins.

- Structural proteins, e.g., actin and tubulin, which are globular and soluble as monomers, but polymerize to form long, stiff fibers.

Members

Among the most known globular proteins is hemoglobin, a member of the globin protein family. Other globular proteins are the alpha, beta and gamma(IgA, IgD, IgE, IgG and IgM) globulin. Nearly all enzymes with major metabolic functions are globular in shape, as well as many signal transduction proteins.

Albumins are also globular proteins, although, unlike all of the other globular proteins, they are completely soluble in water. They are not soluble in oil.

PROTEIN STRUCTURE

Increasingly, drug developers are looking to large molecules and particularly proteins as a therapeutic option. Formulation of a protein drug product can be quite a challenge, but without a good understanding of the nature of protein structure and the conformational characteristics of the specific protein being formulated, the results can be ruinous.

The term structure when used in relation to proteins, takes on a much more complex meaning than it does for small molecules. Proteins are macromolecules and have four different levels of structure – primary, secondary, tertiary and quaternary.

Primary Structure

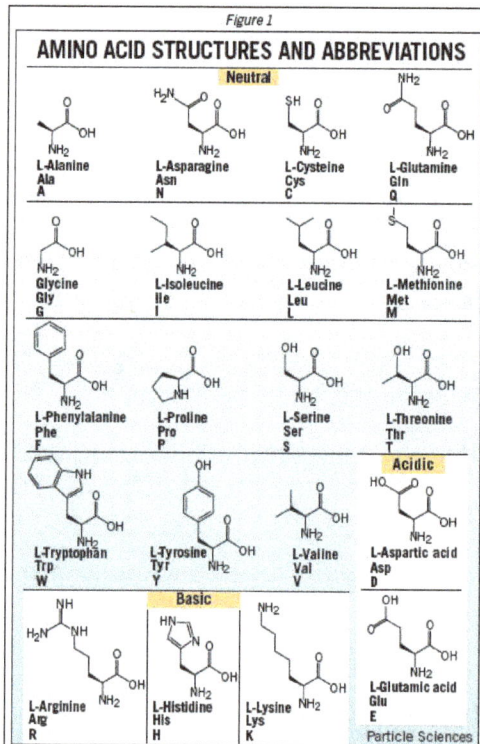

Figure 1

AMINO ACID STRUCTURES AND ABBREVIATIONS

There are 20 different standard L-α-amino acids used by cells for protein construction. Amino acids, as their name indicates, contain both a basic amino group and an acidic carboxyl group. This difunctionality allows the individual amino acids to join together in long chains by forming peptide bonds: amide bonds between the -NH$_2$ of one amino acid and the -COOH of another. Sequences with fewer than 50 amino acids are generally referred to as peptides, while the terms protein or polypeptide are used for longer sequences. A protein can be made up of one or more polypeptide molecules. The end of the peptide or protein sequence with a free carboxyl group is called the carboxy-terminus or C-terminus. The terms amino-terminus or N-terminus describe the end of the sequence with a free α-amino group.

The amino acids differ in structure by the substituent on their side chains. These side chains confer different chemical, physical and structural properties to the final peptide or protein. The structures of the 20 amino acids commonly found in proteins are shown in Figure. Each amino acid has both a one-letter and three-letter abbreviation. These abbreviations are commonly used to simplify the written sequence of a peptide or protein.

Depending on the side-chain substituent, an amino acid can be classified as being acidic, basic or neutral. Although 20 amino acids are required for synthesis of various proteins found in humans, we can synthesize only 10. The remaining 10 are called essential amino acids and must be obtained in the diet.

The amino acid sequence of a protein is encoded in DNA. Proteins are synthesized by a series of steps called transcription (the use of a DNA strand to make a complimentary messenger RNA strand - mRNA) and translation (the mRNA sequence is used as a template to guide the synthesis of the chain of amino acids which make up the protein). Often, post-translational modifications, such as glycosylation or phosphorylation, occur which are necessary for the biological function of the protein. While the amino acid sequence makes up the primary structure of the protein, the chemical/biological properties of the protein are very much dependent on the three-dimensional or tertiary structure.

Secondary Structure

Stretches or strands of proteins or peptides have distinct characteristic local structural conformations or secondary structure, dependent on hydrogen bonding. The two main types of secondary structure are the α-helix and the ß-sheet.

The α-helix is a right-handed coiled strand. The side-chain substituents of the amino acid groups in an α-helix extend to the outside. Hydrogen bonds form between the oxygen of the C=O of each peptide bond in the strand and the hydrogen of the N-H group of the peptide bond four amino acids below it in the helix. The hydrogen bonds make this structure especially stable. The side-chain substituents of the amino acids fit in beside the N-H groups.

The hydrogen bonding in a ß-sheet is between strands (inter-strand) rather than within strands (intra-strand). The sheet conformation consists of pairs of strands lying side-by-side. The carbonyl oxygens in one strand hydrogen bond with the amino hydrogens of the adjacent strand. The two strands can be either parallel or anti-parallel depending on whether the strand directions (N-terminus to C-terminus) are the same or opposite. The anti-parallel ß-sheet is more stable due to the more well-aligned hydrogen bonds.

Tertiary Structure

The overall three-dimensional shape of an entire protein molecule is the tertiary structure. The protein molecule will bend and twist in such a way as to achieve maximum stability or lowest energy state. Although the three-dimensional shape of a protein may seem irregular and random, it is fashioned by many stabilizing forces due to bonding interactions between the side-chain groups of the amino acids.

Under physiologic conditions, the hydrophobic side-chains of neutral, non-polar amino acids such as phenylalanine or isoleucine tend to be buried on the interior of the protein molecule thereby shielding them from the aqueous medium. The alkyl groups of alanine, valine, leucine and isoleucine often form hydrophobic interactions between one-another, while aromatic groups such as those of phenylalanine and tryosine often stack together. Acidic or basic amino acid side-chains will generally be exposed on the surface of the protein as they are hydrophilic.

The formation of disulfide bridges by oxidation of the sulfhydryl groups on cysteine is an important aspect of the stabilization of protein tertiary structure, allowing different parts of the protein chain to be held together covalently. Additionally, hydrogen bonds may form between different side-chain groups. As with disulfide bridges, these hydrogen bonds can bring together two parts of a chain that are some distance away in terms of sequence. Salt bridges, ionic interactions between positively and negatively charged sites on amino acid side chains, also help to stabilize the tertiary structure of a protein.

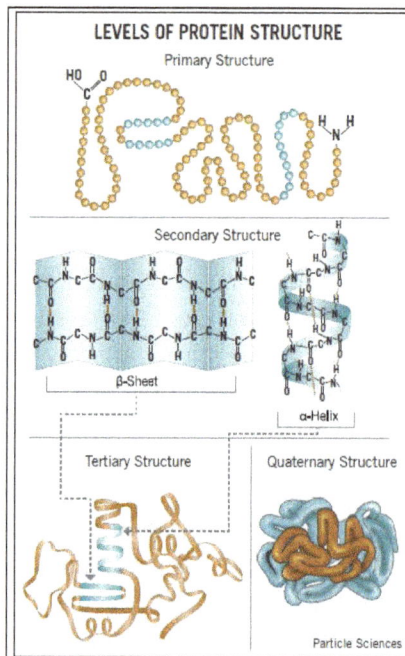

LEVELS OF PROTEIN STRUCTURE

Primary Structure

Secondary Structure

β-Sheet

α-Helix

Tertiary Structure

Quaternary Structure

Particle Sciences

Quaternary Structure

Many proteins are made up of multiple polypeptide chains, often referred to as protein subunits. These subunits may be the same (as in a homodimer) or different (as in a heterodimer). The quaternary structure refers to how these protein subunits interact with each other and arrange themselves to form a larger aggregate protein complex. The final shape of the protein complex is once again stabilized by various interactions, including hydrogen-bonding, disulfide-bridges and salt bridges. The four levels of protein structure.

Protein Stability

Due to the nature of the weak interactions controlling the three-dimensional structure, proteins are very sensitive molecules. The term native state is used to describe the protein in its most stable natural conformation in situ. This native state can be disrupted by a number of external stress factors including temperature, pH, removal of water, presence of hydrophobic surfaces, presence of metal ions and high shear. The loss of secondary, tertiary or quaternary structure due to exposure to a stress factor is called denaturation. Denaturation results in unfolding of the protein into a random or mis-folded shape.

A denatured protein can have quite a different activity profile than the protein in its native form, usually losing biological function. In addition to becoming denatured, proteins can also form aggregates under certain stress conditions. Aggregates are often produced during the manufacturing process and are typically undesirable, largely due to the possibility of them causing adverse immune responses when administered.

In addition to these physical forms of protein degradation, it is also important to be aware of the possible pathways of protein chemical degradation. These include oxidation, deamidation, peptide-bond hydrolysis, disulfide-bond reshuffling and cross-linking. The methods used in the processing and the formulation of proteins, including any lyophilization step, must be carefully examined to prevent degradation and to increase the stability of the protein biopharmaceutical both in storage and during drug delivery.

Protein Structure Analysis

The complexities of protein structure make the elucidation of a complete protein structure extremely difficult even with the most advanced analytical equipment. An amino acid analyzer can be used to determine which amino acids are present and the molar ratios of each. The sequence of the protein can then be analyzed by means of peptide mapping and the use of Edman degradation or mass spectroscopy. This process is routine for peptides and small proteins, but becomes more complex for large multimeric proteins.

Peptide mapping generally entails treatment of the protein with different protease enzymes in order to chop up the sequence into smaller peptides at specific cleavage sites. Two commonly used enzymes are trypsin and chymotrypsin. Mass spectroscopy has become an invaluable tool for the analysis of enzyme digested proteins, by means of peptide fingerprinting methods and database searching. Edman degradation involves the cleavage, separation and identification of one amino acid at a time from a short peptide, starting from the N-terminus.

One method used to characterize the secondary structure of a protein is circular dichroism spectroscopy (CD). The different types of secondary structure, α-helix, ß-sheet and random coil, all have characteristic circular dichroism spectra in the far-uv region of the spectrum (190-250 nm). These spectra can be used to approximate the fraction of the entire protein made up of each type of structure.

A more complete, high-resolution analysis of the three-dimensional structure of a protein is carried out using X-ray crystallography or nuclear magnetic resonance (NMR) analysis. To determine the three-dimensional structure of a protein by X-ray diffraction, a large, well-ordered single crystal is required. X-ray diffraction allows measurement of the short distances between atoms and yields a three-dimensional electron density map, which can be used to build a model of the protein structure.

The use of NMR to determine the three-dimensional structure of a protein has some advantages over X-ray diffraction in that it can be carried out in solution and thus the protein is free of the constraints of the crystal lattice. The two-dimensional NMR techniques generally used are NOESY, which measures the distances between atoms through space, and COESY, which measures distances through bonds.

Protein Structure Stability Analysis

Many different techniques can be used to determine the stability of a protein. For the analysis of unfolding of a protein, spectroscopic methods such as fluorescence, UV, infrared and CD can be used. Thermodynamic methods such as differential scanning calorimetry (DSC) can be useful in determining the effect of temperature on protein stability. Comparative peptide-mapping (usually using LC/MS) is an extremely valuable tool in determining chemical changes in a protein such as oxidation or deamidation. HPLC is also an invaluable means of analyzing the purity of a protein. Other analytical methods such as SDS-PAGE, iso-electric focusing and capillary electrophoresis can also be used to determine protein stability, and a suitable bioassay should be used to determine the potency of a protein biopharmaceutical. The state of aggregation can be determined by following "particle" size and arrayed instruments are now available to follow this over time under various conditions.

The variety of methods for determining protein stability again emphasizes the complexity of the nature of protein structure and the importance of maintaining that structure for a successful biopharmaceutical product.

Protein Structure Prediction

Protein structure prediction is the inference of the three-dimensional structure of a protein from its amino acid sequence—that is, the prediction of its folding and its secondary and tertiary structure from its primary structure. Structure prediction is fundamentally different from the inverse problem of protein design. Protein structure prediction is one of the most important goals pursued by bioinformatics and theoretical chemistry; it is highly important in medicine (for example, in drug design) and biotechnology (for example, in the design of novel enzymes). Every two years, the performance of current methods is assessed in the CASP experiment (Critical Assessment of Techniques for Protein Structure Prediction). A continuous evaluation of protein structure prediction web servers is performed by the community project CAMEO3D.

Protein Structure and Terminology

Proteins are chains of amino acids joined together by peptide bonds. Many conformations of this chain are possible due to the rotation of the chain about each $C\alpha$ atom. It is these conformational changes that are responsible for differences in the three dimensional structure of proteins. Each amino acid in the chain is polar, i.e. it has separated positive and negative charged regions with a free carbonyl group, which can act as hydrogen bond acceptor and an NH group, which can act as hydrogen bond donor. These groups can therefore interact in the protein structure. The 20 amino acids can be classified according to the chemistry of the side chain which also plays an important structural role. Glycine takes on a special position, as it has the smallest side chain, only one hydrogen atom, and therefore can increase the local flexibility in the protein structure. Cysteine on the other hand can react with another cysteine residue and thereby form a cross link stabilizing the whole structure.

The protein structure can be considered as a sequence of secondary structure elements, such as α helices and β sheets, which together constitute the overall three-dimensional configuration of the protein chain. In these secondary structures regular patterns of H bonds are formed between neighboring amino acids, and the amino acids have similar Φ and Ψ angles.

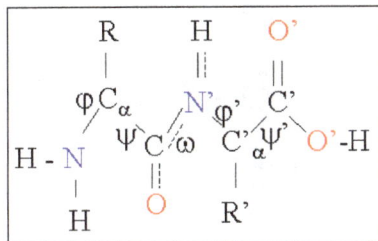

Bond angles for ψ and ω.

The formation of these structures neutralizes the polar groups on each amino acid. The secondary structures are tightly packed in the protein core in a hydrophobic environment. Each amino acid side group has a limited volume to occupy and a limited number of possible interactions with other nearby side chains, a situation that must be taken into account in molecular modeling and alignments.

α Helix

The α helix is the most abundant type of secondary structure in proteins. The α helix has 3.6 amino acids per turn with an H bond formed between every fourth residue; the average length is 10 amino acids (3 turns) or 10 Å but varies from 5 to 40 (1.5 to 11 turns). The alignment of the H bonds creates a dipole moment for the helix with a resulting partial positive charge at the amino end of the helix. Because this region has free NH2 groups, it will interact with negatively charged groups such as phosphates. The most common location of α helices is at the surface of protein cores, where they provide an interface with the aqueous environment. The inner-facing side of the helix tends to have hydrophobic amino acids and the outer-facing side hydrophilic amino acids. Thus, every third of four amino acids along the chain will tend to be hydrophobic, a pattern that can be quite readily detected. In the leucine zipper motif, a repeating pattern of leucines on the facing sides of two adjacent helices is highly predictive of the motif. A helical-wheel plot can be used to show this repeated pattern. Other α helices buried in the protein core or in cellular membranes have a higher and more regular distribution of hydrophobic amino acids, and are highly predictive of such structures. Helices exposed on the surface have a lower proportion of hydrophobic amino acids. Amino acid content can be predictive of an α -helical region. Regions richer in alanine (A), glutamic acid (E), leucine (L), and methionine (M) and poorer in proline (P), glycine (G), tyrosine (Y), and serine (S) tend to form an α helix. Proline destabilizes or breaks an α helix but can be present in longer helices, forming a bend.

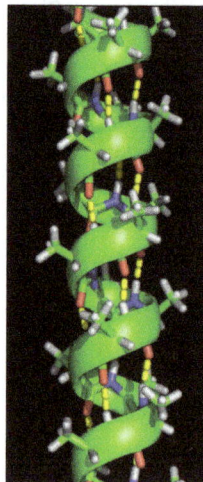

An alpha-helix with hydrogen bonds (yellow dots).

β Sheet

β sheets are formed by H bonds between an average of 5–10 consecutive amino acids in one portion of the chain with another 5–10 farther down the chain. The interacting regions may be adjacent, with a short loop in between, or far apart, with other structures in between. Every chain may run in the same direction to form a parallel sheet, every other chain may run in the reverse chemical direction to form an anti parallel sheet, or the chains may be parallel and anti parallel to form a mixed sheet. The pattern of H bonding is different in the parallel and anti parallel configurations. Each amino acid in the interior strands of the sheet forms two H bonds with neighboring amino acids, whereas each amino acid on the outside strands forms only one bond with an interior strand. Looking across the sheet at right angles to the strands, more distant strands are rotated slightly counterclockwise to form a left-handed twist. The $C\alpha$ atoms alternate above and below the sheet in a pleated structure, and the R side groups of the amino acids alternate above and below the pleats. The Φ and Ψ angles of the amino acids in sheets vary considerably in one region of the Ramachandran plot. It is more difficult to predict the location of β sheets than of α helices. The situation improves somewhat when the amino acid variation in multiple sequence alignments is taken into account.

Loop

Loops are regions of a protein chain that are 1) between α helices and β sheets, 2) of various lengths and three-dimensional configurations, and 3) on the surface of the structure.

Hairpin loops that represent a complete turn in the polypeptide chain joining two anti-parallel β strands may be as short as two amino acids in length. Loops interact with the surrounding aqueous environment and other proteins. Because amino acids in loops are not constrained by space and environment as are amino acids in the core region, and do not have an effect on the arrangement of secondary structures in the core, more substitutions, insertions, and deletions may occur. Thus, in a sequence alignment, the presence of these features may be an indication of a loop. The positions of introns in genomic DNA sometimes correspond to the locations of loops in the encoded protein. Loops also tend to have charged and polar amino acids and are frequently a component of active sites. A detailed examination of loop structures has shown that they fall into distinct families.

Coils

A region of secondary structure that is not an α helix, a β sheet, or a recognizable turn is commonly referred to as a coil.

Protein Classification

Proteins may be classified according to both structural and sequence similarity. For structural classification, the sizes and spatial arrangements of secondary structures described in the above paragraph are compared in known three-dimensional structures. Classification based on sequence similarity was historically the first to be used. Initially, similarity based on alignments of whole sequences was performed. Later, proteins were classified on the basis of the occurrence of conserved amino acid patterns. Databases that classify proteins by one or more of these schemes are available. In considering protein classification schemes, it is important to keep several observations in mind. First, two entirely different protein sequences from different evolutionary origins may fold into a similar structure. Conversely, the sequence of an ancient gene for a given structure may have diverged considerably in different species while at the same time maintaining the same basic structural features. Recognizing any remaining sequence similarity in such cases may be a very difficult task. Second, two proteins that share a significant degree of sequence similarity either with each other or with a third sequence also share an evolutionary origin and should share some structural features also. However, gene duplication and genetic rearrangements during evolution may give rise to new gene copies, which can then evolve into proteins with new function and structure.

Terms used for Classifying Protein Structures and Sequences

The more commonly used terms for evolutionary and structural relationships among proteins are listed below. Many additional terms are used for various kinds of structural features found in proteins. Descriptions of such terms may be found at the CATH Web site, the Structural Classification of Proteins (SCOP) Web site, and a Glaxo-Wellcome tutorial on the Swiss bioinformatics Expasy Web site.

Active site: a localized combination of amino acid side groups within the tertiary (three-dimensional) or quaternary (protein subunit) structure that can interact with a chemically specific substrate and that provides the protein with biological activity. Proteins of very different amino acid sequences may fold into a structure that produces the same active site. Architecture is the relative orientations of secondary structures in a three-dimensional structure without regard to whether or not they share a similar loop structure. Fold (topology) a type of architecture that also has a conserved loop structure.

Blocks: is a conserved amino acid sequence pattern in a family of proteins. The pattern includes a series of possible matches at each position in the represented sequences, but there are not any inserted or deleted positions in the pattern or in the sequences. By way of contrast, sequence profiles are a type of scoring matrix that represents a similar set of patterns that includes insertions and deletions.

Class: a term used to classify protein domains according to their secondary structural content and organization. Four classes were originally recognized by Levitt and Chothia and several others have been added in the SCOP database. Three classes are given in the CATH database: mainly-α, mainly-β, and α–β, with the α–β class including both alternating α/β and $\alpha+\beta$ structures.

Core: the portion of a folded protein molecule that comprises the hydrophobic interior of α-helices and β-sheets. The compact structure brings together side groups of amino acids into close enough proximity so that they can interact. When comparing protein structures, as in the SCOP database, core is the region common to most of the structures that share a common fold or that are in the same superfamily. In structure prediction, core is sometimes defined as the arrangement of secondary structures that is likely to be conserved during evolutionary change.

Domain (sequence context): a segment of a polypeptide chain that can fold into a three-dimensional structure irrespective of the presence of other segments of the chain. The separate domains of a given protein may interact extensively or may be joined only by a length of polypeptide chain. A protein with several domains may use these domains for functional interactions with different molecules.

Family (sequence context): a group of proteins of similar biochemical function that are more than 50% identical when aligned. This same cutoff is still used by the Protein Information Resource (PIR). A protein family comprises proteins with the same function in different organisms (orthologous sequences) but may also include proteins in the same organism (paralogous sequences) derived from gene duplication and rearrangements. If a multiple sequence alignment of a protein family reveals a common level of similarity throughout the lengths of the proteins, PIR refers to the family as a homeomorphic family. The aligned region is referred to as a homeomorphic domain, and this region may comprise several smaller homology domains that are shared with other families. Families may be further subdivided into subfamilies or grouped into superfamilies based on respective higher or lower levels of sequence similarity. The SCOP database reports 1296 families and the CATH database (version 1.7 beta), reports 1846 families.

When the sequences of proteins with the same function are examined in greater detail, some are found to share high sequence similarity. They are obviously members of the same family by the above criteria. However, others are found that have very little, or even insignificant, sequence similarity with other family members. In such cases, the family relationship between two distant family members A and C can often be demonstrated by finding an additional family member B that shares significant similarity with both A and C. Thus, B provides a connecting link between A and C. Another approach is to examine distant alignments for highly conserved matches.

At a level of identity of 50%, proteins are likely to have the same three-dimensional structure, and the identical atoms in the sequence alignment will also superimpose within approximately 1 Å in the structural model. Thus, if the structure of one member of a family is known, a reliable prediction may be made for a second member of the family, and the higher the identity level, the more reliable the prediction. Protein structural modeling can be performed by examining how well the amino acid substitutions fit into the core of the three-dimensional structure.

Family (structural context) as used in the FSSP database (Families of structurally similar proteins) and the DALI/FSSP Web site, two structures that have a significant level of structural similarity but not necessarily significant sequence similarity.

Fold similar to structural motif, includes a larger combination of secondary structural units in the same configuration. Thus, proteins sharing the same fold have the same combination of secondary structures that are connected by similar loops. An example is the Rossman fold comprising several alternating α helices and parallel β strands. In the SCOP, CATH, and FSSP databases, the known protein structures have been classified into hierarchical levels of structural complexity with the fold as a basic level of classification.

Homologous domain (sequence context) an extended sequence pattern, generally found by sequence alignment methods, that indicates a common evolutionary origin among the aligned sequences. A homology domain is generally longer than motifs. The domain may include all of a given protein sequence or only a portion of the sequence. Some domains are complex and made up of several smaller homology domains that became joined to form a larger one during evolution. A domain that covers an entire sequence is called the homeomorphic domain by PIR (Protein Information Resource).

Module a region of conserved amino acid patterns comprising one or more motifs and considered to be a fundamental unit of structure or function. The presence of a module has also been used to classify proteins into families.

Motif (sequence context) a conserved pattern of amino acids that is found in two or more proteins. In the Prosite catalog, a motif is an amino acid pattern that is found in a group of proteins that have a similar biochemical activity, and that often is near the active site of the protein. Examples of sequence motif databases are the Prosite catalog and the Stanford Motifs Database.

Motif (structural context) a combination of several secondary structural elements produced by the folding of adjacent sections of the polypeptide chain into a specific three-dimensional configuration. An example is the helix-loop-helix motif. Structural motifs are also referred to as supersecondary structures and folds.

Position-specific scoring matrix (sequence context, also known as weight or scoring matrix) represents a conserved region in a multiple sequence alignment with no gaps. Each matrix column represents the variation found in one column of the multiple sequence alignment.

Position-specific scoring matrix—3D (structural context) represents the amino acid variation found in an alignment of proteins that fall into the same structural class. Matrix columns represent the amino acid variation found at one amino acid position in the aligned structures.

Primary structure the linear amino acid sequence of a protein, which chemically is a polypeptide chain composed of amino acids joined by peptide bonds.

Profile (sequence context) a scoring matrix that represents a multiple sequence alignment of a protein family. The profile is usually obtained from a well-conserved region in a multiple sequence alignment. The profile is in the form of a matrix with each column representing a position in the alignment and each row one of the amino acids. Matrix values give the likelihood of each amino acid at the corresponding position in the alignment. The profile is moved along the target sequence to locate the best scoring regions by a dynamic programming algorithm. Gaps are allowed during matching and a gap penalty is included in this case as a negative score when no amino acid is matched. A sequence profile may also be represented by a hidden Markov model, referred to as a profile HMM.

Profile (structural context) a scoring matrix that represents which amino acids should fit well and which should fit poorly at sequential positions in a known protein structure. Profile columns represent sequential positions in the structure, and profile rows represent the 20 amino acids. As with a sequence profile, the structural profile is moved along a target sequence to find the highest possible alignment score by a dynamic programming algorithm. Gaps may be included and receive a penalty. The resulting score provides an indication as to whether or not the target protein might adopt such a structure.

Quaternary structure the three-dimensional configuration of a protein molecule comprising several independent polypeptide chains.

Secondary structure the interactions that occur between the C, O, and NH groups on amino acids in a polypeptide chain to form α-helices, β-sheets, turns, loops, and other forms, and that facilitate the folding into a three-dimensional structure.

Superfamily a group of protein families of the same or different lengths that are related by distant yet detectable sequence similarity. Members of a given superfamily thus have a common evolutionary origin. Originally, Dayhoff defined the cutoff for superfamily status as being the chance that the sequences are not related of 10 6, on the basis of an alignment score (Dayhoff et al. 1978). Proteins with few identities in an alignment of the sequences but with a convincingly common number of structural and functional features are placed in the same superfamily. At the level of three-dimensional structure, superfamily proteins will share common structural features such as a common fold, but there may also be differences in the number and arrangement of secondary structures. The PIR resource uses the term *homeomorphic superfamilies* to refer to superfamilies that are composed of sequences that can be aligned from end to end, representing a sharing of single sequence homology domain, a region of similarity that extends throughout the alignment. This domain may also comprise smaller homology domains that are shared with other protein families and superfamilies. Although a given protein sequence may contain domains found in several superfamilies, thus indicating a complex evolutionary history, sequences will be assigned to only one homeomorphic superfamily based on the presence of similarity throughout a multiple sequence alignment. The superfamily alignment may also include regions that do not align either within or at the ends of the alignment. In contrast, sequences in the same family align well throughout the alignment.

Supersecondary structure a term with similar meaning to a structural motif. Tertiary structure is the three-dimensional or globular structure formed by the packing together or folding of secondary structures of a polypeptide chain.

Secondary Structure

Secondary structure prediction is a set of techniques in bioinformatics that aim to predict the local secondary structures of proteins based only on knowledge of their amino acid sequence. For proteins, a prediction consists of assigning regions of the amino acid sequence as likely alpha helices, beta strands (often noted as "extended" conformations), or turns. The success of a prediction is determined by comparing it to the results of the DSSP algorithm (or similar e.g. STRIDE) applied to the crystal structure of the protein. Specialized algorithms have been developed for the detection of specific well-defined patterns such as transmembrane helices and coiled coils in proteins.

The best modern methods of secondary structure prediction in proteins reach about 80% accuracy; this high accuracy allows the use of the predictions as feature improving fold recognition and ab initio protein structure prediction, classification of structural motifs, and refinement of sequence alignments. The accuracy of current protein secondary structure prediction methods is assessed in weekly benchmarks such as LiveBench and EVA.

Early methods of secondary structure prediction, introduced in the 1960s and early 1970s, focused on identifying likely alpha helices and were based mainly on helix-coil transition models. Significantly more accurate predictions that included beta sheets were introduced in the 1970s and relied on statistical assessments based on probability parameters derived from known solved structures. These methods, applied to a single sequence, are typically at most about 60-65% accurate, and often underpredict beta sheets. The evolutionary conservation of secondary structures can be exploited by simultaneously assessing many homologous sequences in a multiple sequence alignment, by calculating the net secondary structure propensity of an aligned column of amino acids. In concert with larger databases of known protein structures and modern machine learning methods such as neural nets and support vector machines, these methods can achieve up to 80% overall accuracy in globular proteins. The theoretical upper limit of accuracy is around 90%, partly due to idiosyncrasies in DSSP assignment near the ends of secondary structures, where local conformations vary under native conditions but may be forced to assume a single conformation in crystals due to packing constraints. Limitations are also imposed by secondary structure prediction's inability to account for tertiary structure; for example, a sequence predicted as a likely helix may still be able to adopt a beta-strand conformation if it is located within a beta-sheet region of the protein and its side chains pack well with their neighbors. Dramatic conformational changes related to the protein's function or environment can also alter local secondary structure.

To date, over 20 different secondary structure prediction methods have been developed. One of the first algorithms was Chou-Fasman method, which relies predominantly on probability parameters determined from relative frequencies of each amino acid's appearance in each type of secondary structure. The original Chou-Fasman parameters, determined from the small sample of structures solved in the mid-1970s, produce poor results compared to modern methods, though the parameterization has been updated since it was first published. The Chou-Fasman method is roughly 50-60% accurate in predicting secondary structures.

The next notable program was the GOR method, named for the three scientists who developed it — *G*arnier, *O*sguthorpe, and *R*obson, is an information theory-based method. It uses the more powerful probabilistic technique of Bayesian inference. The GOR method takes into account not only the probability of each amino acid having a particular secondary structure, but also the conditional probability of the amino acid assuming each structure given the contributions of its neighbors (it does not assume that the neighbors have that same structure). The approach is both more sensitive and more accurate than that of Chou and Fasman because amino acid structural propensities are only strong for a small number of amino acids such as proline and glycine. Weak contributions from each of many neighbors can add up to strong effects overall. The original GOR method was roughly 65% accurate and is dramatically more successful in predicting alpha helices than beta sheets, which it frequently mispredicted as loops or disorganized regions.

Another big step forward, was using machine learning methods. First artificial neural networks methods were used. As a training sets they use solved structures to identify common sequence motifs associated with particular arrangements of secondary structures. These methods are over 70% accurate in their predictions, although beta strands are still often underpredicted due to the lack of three-dimensional structural information that would allow assessment of hydrogen bonding patterns that can promote formation of the extended conformation required for the presence of a complete beta sheet. PSIPRED and JPRED are some of the most known programs based on neural networks for protein secondary structure prediction. Next, support vector machines have proven particularly useful for predicting the locations of turns, which are difficult to identify with statistical methods.

Extensions of machine learning techniques attempt to predict more fine-grained local properties of proteins, such as backbone dihedral angles in unassigned regions. Both SVMs and neural networks have been applied to this problem. More recently, real-value torsion angles can be accurately predicted by SPINE-X and successfully employed for ab initio structure prediction.

Other Improvements

It is reported that in addition to the protein sequence, secondary structure formation depends on other factors. For example, it is reported that secondary structure tendencies depend also on local environment, solvent accessibility of residues, protein structural class, and even the organism from which the proteins are obtained. Based on such observations, some studies have shown that secondary structure prediction can be improved by addition of information about protein structural class, residue accessible surface area and also contact number information.

Tertiary Structure

The practical role of protein structure prediction is now more important than ever. Massive amounts of protein sequence data are produced by modern large-scale DNA sequencing efforts such as the Human Genome Project. Despite community-wide efforts in structural genomics, the output of experimentally determined protein structures—typically by time-consuming and relatively expensive X-ray crystallography or NMR spectroscopy—is lagging far behind the output of protein sequences.

The protein structure prediction remains an extremely difficult and unresolved undertaking. The two main problems are calculation of protein free energy and finding the global minimum of this energy. A protein structure prediction method must explore the space of possible protein structures which is astronomically large. These problems can be partially bypassed in "comparative" or homology modeling and fold recognition methods, in which the search space is pruned by the assumption that the protein in question adopts a structure that is close to the experimentally determined structure of another homologous protein. On the other hand, the *de novo* or ab initio protein structure prediction methods must explicitly resolve these problems. The progress and challenges in protein structure prediction has been reviewed in Zhang 2008.

Before Modelling

Most tertiary structure modelling methods, such as Rosetta, are optimized for modelling the tertiary structure of single protein domains. A step called domain parsing, or domain boundary prediction, is usually done first to split a protein into potential structural domains. As with the rest of tertiary structure prediction, this can be done comparatively from known structures or *ab initio* with the sequence only (usually by machine learning, assisted by covariation). The structures for individual domains are docked together in a process called domain assembly to form the final tertiary structure.

Ab Initio Protein Modelling

Energy and Fragment-based Methods

Ab initio- or de novo- protein modelling methods seek to build three-dimensional protein models "from scratch", i.e., based on physical principles rather than (directly) on previously solved structures. There are many possible procedures that either attempt to mimic protein folding or apply some stochastic method to search possible solutions (i.e., global optimization of a suitable energy function). These procedures tend to require vast computational resources, and have thus only been carried out for tiny proteins. To predict protein structure de novo for larger proteins will require better algorithms and larger computational resources like those afforded by either powerful supercomputers (such as Blue Gene or MDGRAPE-3) or distributed computing. Although these computational barriers are vast, the potential benefits of structural genomics (by predicted or experimental methods) make ab initio structure prediction an active research field.

As of 2009, a 50-residue protein could be simulated atom-by-atom on a supercomputer for 1 millisecond. As of 2012, comparable stable-state sampling could be done on a standard desktop with a new graphics card and more sophisticated algorithms. A much larger simulation timescales can be achieved using coarse-grained modeling.

Evolutionary Covariation to Predict 3D Contacts

As sequencing became more commonplace in the 1990s several groups used protein sequence alignments to predict correlated mutations and it was hoped that these co-evolved residues could be used to predict tertiary structure (using the analogy to distance constraints from experimental procedures such as NMR). The assumption is when single residue mutations are slightly deleterious, compensatory mutations may occur to restabilize residue-residue interactions. This early work used what are known as *local* methods to calculate correlated mutations from protein sequences, but suffered from indirect false correlations which result from treating each pair of residues as independent of all other pairs.

In 2011, a different, and this time *global* statistical approach, demonstrated that predicted coevolved residues were sufficient to predict the 3D fold of a protein, providing there are enough sequences available (>1,000 homologous sequences are needed). The method, EVfold, uses no homology modeling, threading or 3D structure fragments and can be run on a standard personal computer even for proteins with hundreds of residues. The accuracy of the contacts predicted using this and related approaches has now been demonstrated on many known structures and contact maps, including the prediction of experimentally unsolved transmembrane proteins.

Comparative Protein Modeling

Comparative protein modelling uses previously solved structures as starting points, or templates. This is effective because it appears that although the number of actual proteins is vast, there is a limited set of tertiary structural motifs to which most proteins belong. It has been suggested that there are only around 2,000 distinct protein folds in nature, though there are many millions of different proteins.

These methods may also be split into two groups:

- Homology modeling is based on the reasonable assumption that two homologous proteins will share very similar structures. Because a protein's fold is more evolutionarily conserved than its amino acid sequence, a target sequence can be modeled with reasonable accuracy on a very distantly related template, provided that the relationship between target and template can be discerned through sequence alignment. It has been suggested that the primary bottleneck in comparative modelling arises from difficulties in alignment rather than from errors in structure prediction given a known-good alignment. Unsurprisingly, homology modelling is most accurate when the target and template have similar sequences.

- Protein threading scans the amino acid sequence of an unknown structure against a database of solved structures. In each case, a scoring function is used to assess the compatibility of the sequence to the structure, thus yielding possible three-dimensional models. This type of method is also known as 3D-1D fold recognition due to its compatibility analysis between three-dimensional structures and linear protein sequences. This method has also given rise to methods performing an inverse folding search by evaluating the compatibility of a given structure with a large database of sequences, thus predicting which sequences have the potential to produce a given fold.

Side-chain Geometry Prediction

Accurate packing of the amino acid side chains represents a separate problem in protein structure prediction. Methods that specifically address the problem of predicting side-chain geometry include dead-end elimination and the self-consistent mean field methods. The side chain conformations with low energy are usually determined on the rigid polypeptide backbone and using a set of discrete side chain conformations known as "rotamers." The methods attempt to identify the set of rotamers that minimize the model's overall energy.

These methods use rotamer libraries, which are collections of favorable conformations for each residue type in proteins. Rotamer libraries may contain information about the conformation, its frequency, and the standard deviations about mean dihedral angles, which can be used in sampling. Rotamer libraries are derived from structural bioinformatics or other statistical analysis of side-chain conformations in known experimental structures of proteins, such as by clustering the observed conformations for tetrahedral carbons near the staggered (60°, 180°, -60°) values.

Rotamer libraries can be backbone-independent, secondary-structure-dependent, or backbone-dependent. Backbone-independent rotamer libraries make no reference to backbone conformation, and are calculated from all available side chains of a certain type (for instance, the first example of a rotamer library, done by Ponder and Richards at Yale in 1987). Secondary-structure-dependent libraries present different dihedral angles and/or rotamer frequencies for α-helix, β-sheet, or coil secondary structures. Backbone-dependent rotamer libraries present conformations and frequencies dependent on the local backbone conformation as defined by the backbone dihedral angles ϕ and ψ, regardless of secondary structure.

The modern versions of these libraries as used in most software are presented as multidimensional distributions of probability or frequency, where the peaks correspond to the dihedral-angle conformations considered as individual rotamers in the lists. Some versions are based on very carefully curated data and are used primarily for structure validation, while others emphasize relative frequencies in much larger data sets and are the form used primarily for structure prediction, such as the Dunbrack rotamer libraries.

Side-chain packing methods are most useful for analyzing the protein's hydrophobic core, where side chains are more closely packed; they have more difficulty addressing the looser constraints and higher flexibility of surface residues, which often occupy multiple rotamer conformations rather than just one.

Prediction of Structural Classes

Statistical methods have been developed for predicting structural classes of proteins based on their amino acid composition, pseudo amino acid composition and functional domain composition. Secondary structure predicion also implicitly generates such a prediction for singular domains.

Quaternary Structure

In the case of complexes of two or more proteins, where the structures of the proteins are known or can be predicted with high accuracy, protein–protein docking methods can be used to predict the structure of the complex. Information of the effect of mutations at specific sites on the affinity of the complex helps to understand the complex structure and to guide docking methods.

PROTEIN PRODUCTION

Protein production is the biotechnological process of generating a specific protein. It is typically achieved by the manipulation of gene expression in an organism such that it expresses large amounts of a recombinant gene. This includes the transcription of the recombinant DNA to messenger RNA (mRNA), the translation of mRNA into polypeptide chains, which are ultimately folded into functional proteins and may be targeted to specific subcellular or extracellular locations.

Protein production systems (in lab jargon also referred to as 'expression systems') are used in the life sciences, biotechnology, and medicine. Molecular biology research uses numerous proteins and enzymes, many of which are from expression systems; particularly DNA polymerase for PCR, reverse transcriptase for RNA analysis, restriction endonucleases for cloning, and to make proteins that are screened in drug discovery as biological targets or as potential drugs themselves. There are also significant applications for expression systems in industrial fermentation, notably the production of biopharmaceuticals such as human insulin to treat diabetes, and to manufacture enzymes.

Protein Production Systems

Commonly used protein production systems include those derived from bacteria, yeast,baculovirus/insect, mammalian cells, and more recently filamentous fungi such as Myceliophthora thermophila. When biopharmaceuticals are produced with one of these systems, process-related impurities termed host cell proteins also arrive in the final product in trace amounts.

Cell-based Systems

The oldest and most widely used expression systems are cell-based and may be defined as the "combination of an expression vector, its cloned DNA, and the host for the vector that provide a context to allow foreign gene function in a host cell, that is, produce proteins at a high level". Overexpression is an abnormally and excessively high level of gene expression which produces a pronounced gene-related phenotype.

There are many ways to introduce foreign DNA to a cell for expression, and many different host cells may be used for expression — each expression system has distinct advantages and liabilities. Expression systems are normally referred to by the host and the DNA source or the delivery mechanism for the genetic material. For example, common hosts are bacteria (such as E.coli, B. subtilis), yeast (such as S.cerevisiae) or eukaryotic cell lines. Common DNA sources and delivery mechanisms are viruses (such as baculovirus, retrovirus, adenovirus), plasmids, artificial chromosomes and bacteriophage (such as lambda). The best expression system depends on the gene involved, for example the Saccharomyces cerevisiae is often preferred for proteins that require significant posttranslational modification. Insect or mammal cell lines are used when human-like splicing of mRNA is required. Nonetheless, bacterial expression has the advantage of easily producing large amounts of protein, which is required for X-ray crystallography or nuclear magnetic resonance experiments for structure determination.

Because bacteria are prokaryotes, they are not equipped with the full enzymatic machinery to accomplish the required post-translational modifications or molecular folding. Hence, multi-domain eukaryotic proteins expressed in bacteria often are non-functional. Also, many proteins become insoluble as inclusion bodies that are difficult to recover without harsh denaturants and subsequent cumbersome protein-refolding.

To address these concerns, expressions systems using multiple eukaryotic cells were developed for applications requiring the proteins be conformed as in, or closer to eukaryotic organisms: cells of plants (i.e. tobacco), of insects or mammalians (i.e. bovines) are transfected with genes and cultured in suspension and even as tissues or whole organisms, to produce fully folded proteins. Mammalian in vivo expression systems have however low yield and other limitations (time-consuming, toxicity to host cells,..). To combine the high yield/productivity and scalable protein features of bacteria and yeast, and advanced epigenetic features of plants, insects and mammalians systems, other protein production systems are developed using unicellular eukaryotes (i.e. non-pathogenic 'Leishmania' cells).

Bacterial Systems

Escherichia Coli

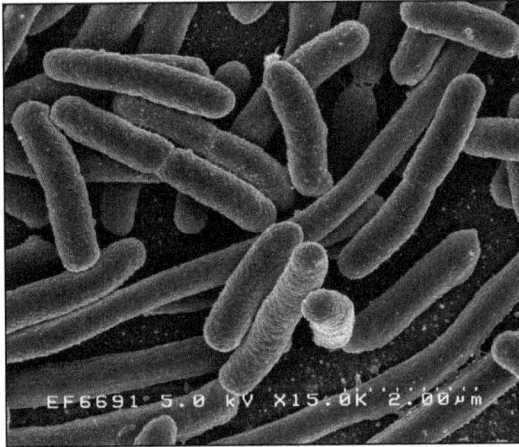

E. coli, one of the most popular hosts for artificial gene expression.

E. coli is one of the most widely used expression hosts, and DNA is normally introduced in a plasmid expression vector. The techniques for overexpression in *E. coli* are well developed and work by increasing the number of copies of the gene or increasing the binding strength of the promoter region so assisting transcription.

For example, a DNA sequence for a protein of interest could be cloned or subcloned into a high copy-number plasmid containing the *lac* (often LacUV5) promoter, which is then transformed into the bacterium *E. coli*. Addition of IPTG (a lactose analog) activates the lac promoter and causes the bacteria to express the protein of interest.

E. coli strain BL21 and BL21(DE3) are two strains commonly used for protein production. As members of the B lineage, they lack lon and OmpT proteases, protecting the produced proteins from degradation. The DE3 prophage found in BL21(DE3) provides T7 RNA polymerase (driven by the LacUV5 promoter), allowing for vectors with the T7 promoter to be used instead.

Corynebacterium

Non-pathogenic species of the gram-positive Corynebacterium are used for the commercial production of various amino acids. The C. glutamicum species is widely used for producing glutamate and lysine, components of human food, animal feed and pharmaceutical products.

Expression of functionally active human epidermal growth factor has been done in C. glutamicum, thus demonstrating a potential for industrial-scale production of human proteins. Expressed proteins can be targeted for secretion through either the general, secretory pathway (Sec) or the twin-arginine translocation pathway (Tat).

Unlike gram-negative bacteria, the gram-positive Corynebacterium lack lipopolysaccharides that function as antigenic endotoxins in humans.

Pseudomonas Fluorescens

The non-pathogenic and gram-negative bacteria, Pseudomonas fluorescens, is used for high level production of recombinant proteins; commonly for the development bio-therapeutics and vaccines. P. fluorescens is a metabolically versatile organism, allowing for high throughput screening and rapid development of complex proteins. P. fluorescens is most well known for its ability to rapid and successfully produce high titers of active, soluble protein.

Eukaryotic Systems

Yeasts

Expression systems using either *S. cerevisiae* or *Pichia pastoris* allow stable and lasting production of proteins that are processed similarly to mammalian cells, at high yield, in chemically defined media of proteins.

Filamentous Fungi

Filamentous fungi, especially Aspergillus and Trichoderma, but also more recently Myceliophthora thermophila C1 have been developed into expression platforms for screening and production of diverse industrial enzymes. The expression system C1 shows a low viscosity morphology in submerged culture, enabling the use of complex growth and production media.

Baculovirus-infected Cells

Baculovirus-infected insect cells (Sf9, Sf21, High Five strains) or mammalian cells (HeLa, HEK 293) allow production of glycosylated or membrane proteins that cannot be produced using fungal or bacterial systems. It is useful for production of proteins in high quantity. Genes are not expressed continuously because infected host cells eventually lyse and die during each infection cycle.

Non-lytic Insect Cell Expression

Non-lytic insect cell expression is an alternative to the lytic baculovirus expression system. In non-lytic expression, vectors are transiently or stably transfected into the chromosomal DNA of insect cells for subsequent gene expression. This is followed by selection and screening of recombinant clones. The non-lytic system has been used to give higher protein yield and quicker expression of recombinant genes compared to baculovirus-infected cell expression. Cell lines used for this system include: Sf9, Sf21 from Spodoptera frugiperda cells, Hi-5 from Trichoplusia ni cells, and Schneider 2 cells and Schneider 3 cells from Drosophila melanogaster cells. With this system, cells do not lyse and several cultivation modes can be used. Additionally, protein production runs are reproducible. This system gives a homogeneous product. A drawback of this system is the requirement of an additional screening step for selecting viable clones.

Excavata

Leishmania tarentolae (cannot infect mammals) expression systems allow stable and lasting production of proteins at high yield, in chemically defined media. Produced proteins exhibit fully eukaryotic post-translational modifications, including glycosylation and disulfide bond formation.

Mammalian Systems

The most common mammalian expression systems are Chinese Hamster ovary (CHO) and Human embryonic kidney (HEK) cells.

- Chinese hamster ovary cell

- Mouse myeloma lymphoblstoid (e.g. NS0 cell)

- Fully Human

  ◦ Human embryonic kidney cells (HEK-293)

  ◦ Human embryonic retinal cells (Crucell's Per.C6)

  ◦ Human amniocyte cells (Glycotope and CEVEC)

Cell-free Systems

Cell-free production of proteins is performed *in vitro* using purified RNA polymerase, ribosomes, tRNA and ribonucleotides. These reagents may be produced by extraction from cells or from a cell-based expression system. Due to the low expression levels and high cost of cell-free systems, cell-based systems are more widely used.

PROTEIN QUANTIFICATION

Protein quantification is necessary to understand the total protein content in a sample or in a formulated product. Accurate protein quantification is important as a range of other critical assays require precise total protein content results in order to generate data. Careful selection of the protein quantification method is required to ensure accurate data. Colorimetric assays, for example, are dependent on the use of an external reference standard and the different absorbance properties between product and the reference standard may lead to inaccuracies. Quantitative amino acid analysis can also be used for protein content determination but an understanding of how different proteins hydrolyze under similar conditions and the stability of amino acids during hydrolysis is critical.

PROTEIN DOMAINS

WW Domain

The WW domain, (also known as the rsp5-domain or WWP repeating motif) is a modular protein domain that mediates specific interactions with protein ligands. This domain is found in a number of unrelated signaling and structural proteins and may be repeated up to four times in some proteins. Apart from binding preferentially to proteins that are proline-rich, with particular proline-motifs, [AP]-P-P-[AP]-Y, some WW domains bind to phosphoserine- phosphothreonine-containing motifs.

Structure and Ligands

The WW domain is one of the smallest protein modules, composed of only 40 amino acids, which mediates specific protein-protein interactions with short proline-rich or proline-containing motifs. Named after the presence of two conserved tryptophans (W), which are spaced 20-22 amino acids apart within the sequence, the WW domain folds into a meandering triple-stranded beta sheet. The identification of the WW domain was facilitated by the analysis of two splice isoforms of YAP gene product, named YAP1-1 and YAP1-2, which differed by the presence of an extra 38 amino acids. These extra amino acids are encoded by a spliced-in exon and represent the second WW domain in YAP1-2 isoform.

The first structure of the WW domain was determined in solution by NMR approach. It represented the WW domain of human YAP in complex with peptide ligand containing Proline-Proline-x–Tyrosine (PPxY where x = any amino acid) consensus motif. Recently, the YAP WW domain structure in complex with SMAD-derived, PPxY motif-containing peptide was further refined. Apart from the PPxY motif, certain WW domains recognize LPxY motif (where L is Leucine), and several WW domains bind to phospho-Serine-Proline (p-SP) or phospho-Threonine-Proline (p-TP) motifs in a phospho-dependent manner. Structures of these WW domain complexes confirmed molecular details of phosphorylation-regulated interactions. There are also WW domains that interact with polyprolines that are flanked by arginine residues or interrupted by leucine residues, but they do not contain aromatic amino acids.

Signaling Function

The WW domain is known to mediate regulatory protein complexes in various signaling networks, including the Hippo signaling pathway. The importance of WW domain-mediated complexes in signaling was underscored by the characterization of genetic syndromes that are caused by loss-of-function point mutations in the WW domain or its cognate ligand. These syndromes are Golabi-Ito-Hall syndrome of intellectual disability caused by missense mutation in a WW domain and Liddle syndrome of hypertension caused by point mutations within PPxY motif.

Examples:

A large variety of proteins containing the WW domain are known. These include; dystrophin, a multidomain cytoskeletal protein; utrophin, a dystrophin-like protein; vertebrate YAP protein, substrate of LATS1 and LATS2 serine-theronine kinases of the Hippo tumor suppressor pathway; *Mus musculus* (Mouse) NEDD4, involved in the embryonic development and differentiation of the central nervous system; *Saccharomyces cerevisiae* (Baker's yeast) RSP5, similar to NEDD4 in its molecular organization; *Rattus norvegicus* (Rat) FE65, a transcription-factor activator expressed preferentially in brain; *Nicotiana tabacum* (Common tobacco) DB10 protein, amongst others.

In 2004, the first comprehensive protein-peptide interaction map for a human modular domain was reported using individually expressed WW domains and genome predicted, PPxY-containing synthetic peptides. At present in the human proteome, 98 WW domains and more than 2000 PPxY-containing peptides, have been identified from sequence analysis of the genome.

Inhibitor

YAP is a WW domain-containing protein that functions as a potent oncogene. Its WW domains must be intact for YAP to act as a transcriptional co-activator that induces expression of proliferative genes. Recent study has shown that endohedral metallofullerenol, a compound that was originally developed as a contrasting agent for MRI (magnetic resonance imaging), has antineoplastic properties. Via molecular dynamic simulations, the ability of this compound to outcompete proline-rich peptides and bind effectively to the WW domain of YAP was documented. Endotheral metallofullerenol may represent a lead compound for the development of therapies for cancer patients who harbor amplified or overexpressed YAP.

Protein Folding

Because of its small size and well-defined structure, the WW domain became a favorite subject of protein folding studies. Among these studies, the work of Rama Ranganathan and David E. Shaw are notable. Ranganathan's team has shown that a simple statistical energy function, which identifies co-evolution between amino acid residues within the WW domain, is necessary and sufficient to specify sequence that folds into native structure. Using such an algorithm, he and his team synthesized libraries of artificial WW domains that functioned in a very similar manner to their natural counterparts, recognizing class-specific proline-rich ligand peptides, The Shaw laboratory developed a specialized machine that allowed elucidation of the atomic level behavior of the WW domain on a biologically relevant time scale. He and his team employed equilibrium simulations of a WW domain and identified seven unfolding and eight folding events that follow the same folding route.

Being a relatively short, 30 to 35 amino acids long, WW domain is amenable to chemical synthesis. It is cooperatively folded and can host chemically introduced non-canonical amino acids. Based on these properties, WW domain has been shown to be a versatile platform for the chemical interrogation of intramolecular interactions and conformational propensities in folded proteins

Pleckstrin Homology Domain

Pleckstrin homology domain (PH domain) is a protein domain of approximately 120 amino acids that occurs in a wide range of proteins involved in intracellular signaling or as constituents of the cytoskeleton.

This domain can bind phosphatidylinositol lipids within biological membranes (such as phosphatidylinositol (3,4,5)-trisphosphate and phosphatidylinositol (4,5)-bisphosphate), and proteins such as the βγ-subunits of heterotrimeric G proteins, and protein kinase C. Through these interactions, PH domains play a role in recruiting proteins to different membranes, thus targeting them to appropriate cellular compartments or enabling them to interact with other components of the signal transduction pathways.

Lipid Binding Specificity

Individual PH domains possess specificities for phosphoinositides phosphorylated at different sites within the inositol ring, e.g., some bind phosphatidylinositol (4,5)-bisphosphate but not phosphatidylinositol (3,4,5)-trisphosphate or phosphatidylinositol (3,4)-bisphosphate, while others may possess the requisite affinity. This is important because it makes the recruitment of different PH domain containing proteins sensitive to the activities of enzymes that either phosphorylate or dephosphorylate these sites on the inositol ring, such as phosphoinositide 3-kinase or PTEN, respectively. Thus, such enzymes exert a part of their effect on cell function by modulating the localization of downstream signaling proteins that possess PH domains that are capable of binding their phospholipid products.

Structure

The 3D structure of several PH domains has been determined. All known cases have a common structure consisting of two perpendicular anti-parallel beta sheets, followed by a C-terminal amphipathic helix. The loops connecting the beta-strands differ greatly in length, making the PH domain relatively difficult to detect while providing the source of the domain's specificity. The only conserved residue among PH domains is a single tryptophan located within the alpha helix that serves to nucleate the core of the domain.

Proteins Containing PH Domain

PH domains can be found in many different proteins, such as OSBP or ARF. Recruitment to the Golgi in this case is dependent on both PtdIns and ARF. A large number of PH domains have poor affinity for phosphoinositides and are hypothesized to function as protein binding domains. A Genome-wide look in *Saccharomyces cerevisiae* showed that most of the 33 yeast PH domains are indeed promiscuous in binding to phosphoinositides, while only one (Num1-PH) behaved highly specific. Proteins reported to contain PH domains belong to the following families:

- Pleckstrin, the protein where this domain was first detected, is the major substrate of protein kinase C in platelets. Pleckstrin contains two PH domains. ARAP proteins contain five PH domains.

- Ser/Thr protein kinases such as the Akt/Rac family, the beta-adrenergic receptor kinases, the mu isoform of PKC and the trypanosomal NrkA family.

- Tyrosine protein kinases belonging to the Btk/Itk/Tec subfamily.

- Insulin receptor substrate 1 (IRS-1).

- Regulators of small G-proteins: 64 RhoGEFs of the Dbl-like family., and several GTPase activating proteins like ABR, BCR or ARAP proteins.

- Cytoskeletal proteins such as dynamin , *Caenorhabditis elegans* kinesin-like protein unc-104, spectrin beta-chain, syntrophin,and S. cerevisiae nuclear migration protein NUM1.

- Oxysterol-binding proteins OSBP, S. cerevisiae OSH1 and YHR073w.

- Ceramide kinase, a lipid kinase that phosphorylates ceramides to ceramide-1-phosphate.

Examples:

Human genes encoding proteins containing this domain include:

- ABR, ADRBK1, ADRBK2, AFAP, AFAP1, AFAP1L1, AFAP1L2, AKAP13, AKT1, AKT2, AKT3, ANLN, APBB1IP, APPL1, APPL2, ARHGAP10, ARHGAP12, ARHGAP15, ARHGAP21, ARHGAP22, ARHGAP23, ARHGAP24, ARHGAP25, ARHGAP26, ARHGAP27, ARHGAP9, ARHGEF16, ARHGEF18, ARHGEF19, ARHGEF2, ARHGEF3, ARHGEF4, ARHGEF5, ARHGEF6, ARHGEF7, ARHGEF9, ASEF2,

- BMX, BTK,

- C20orf42, C9orf100, CADPS, CADPS2, CDC42BPA, CDC42BPB, CDC42BPG, CENTA1, CENTA2, CENTB1, CENTB2, CENTB5, CENTD1, CENTD2, CENTD3, CENTG1, CENTG2, CENTG3, CIT, CNKSR1, CNKSR2, COL4A3BP, CTGLF1, CTGLF2, CTGLF3, CTGLF4, CTGLF5, CTGLF6,

- DAB2IP, DAPP1, DDEF1, DDEF2, DDEFL1, DEF6, DEPDC2, DGKD, DGKH, DGKK, DNM1, DNM2, DNM3, DOCK10, DOCK11, DOCK9, DOK1, DOK2, DOK3, DOK4, DOK5, DOK6, DTGCU2,

- EXOC8,

- FAM109A, FAM109B, FARP1, FARP2, FGD1, FGD2, FGD3, FGD4, FGD5, FGD6,

- GAB1, GAB2, GAB3, GAB4, GRB10, GRB14, GRB7,

- IRS1, IRS2, IRS4, ITK, ITSN1, ITSN2,

- KALRN, KIF1A, KIF1B, KIF1Bbeta,

- MCF2, MCF2L, MCF2L2, MRIP, MYO10,

- NET1, NGEF,

- OBPH1, OBSCN, OPHN1, OSBP, OSBP2, OSBPL10, OSBPL11, OSBPL3, OSBPL5, OSBPL6, OSBPL7, OSBPL8, OSBPL9,

- PHLDA2, PHLDA3, PHLDB1, PHLDB2, PHLPP, PIP3-E, PLCD1, PLCD4, PLCG1, PLCG2, PLCH1, PLCH2, PLCL1, PLCL2, PLD1, PLD2, PLEK, PLEK2, PLEKHA1, PLEKHA2, PLEKHA3, PLEKHA4, PLEKHA5, PLEKHA6, PLEKHA7, PLEKHA8, PLEKHB1, PLEKHB2, PLEKHC1, PLEKHF1, PLEKHF2, PLEKHG1, PLEKHG2, PLEKHG3, PLEKHG4, PLEKHG5, PLEKHG6, PLEKHH1, PLEKHH2, PLEKHH3, PLEKHJ1, PLEKHK1, PLEKHM1, PLEKHM2, PLEKHO1, PLEKHQ1, PREX1, PRKCN, PRKD1, PRKD2, PRKD3, PSCD1, PSCD2, PSCD3, PSCD4, PSD, PSD2, PSD3, PSD4, RALGPS1, RALGPS2, RAPH1,

- RASA1, RASA2, RASA3, RASA4, RASAL1, RASGRF1, RGNEF, ROCK1, ROCK2, RTKN,

- SBF1, SBF2, SCAP2, SGEF, SH2B, SH2B1, SH2B2, SH2B3, SH3BP2, SKAP1, SKAP2, SNTA1, SNTB1, SNTB2, SOS1, SOS2, SPATA13, SPNB4, SPTBN1, SPTBN2, SPTBN4, SPTBN5, STAP1, SWAP70, SYNGAP1,

- TBC1D2, TEC, TIAM1, TRIO, TRIOBP, TYL,

- URP1, URP2,

- VAV1, VAV2, VAV3, VEPH1

Calponin Homology Domain

Calponin homology domain (or CH domain) is a family of actin binding domains found in both cytoskeletal proteins and signal transduction proteins. The domain is about 100 amino acids in length and is composed of four alpha helices. It comprises the following groups of actin-binding domains:

- Actinin-type (including spectrin, fimbrin, ABP-280)

- Calponin-type

The CH domain is involved in actin binding in some members of the family. However, in calponins there is evidence that the CH domain is not involved in its actin binding activity. Most proteins have two copies of the CH domain, however some proteins such as calponin and the human vav proto-oncogene (P15498) have only a single copy. The structure of an example CH domain has been determined using X-ray crystallography.

Examples:

Human genes encoding calponin homology domain-containing proteins include:

- ACTN1, ACTN2, ACTN3, ACTN4, ARHGEF6, ARHGEF7, ASPM,

- CLMN, CNN1, CNN2, CNN3,

- DIXDC1, DMD, DST,

- EHBP1, EHBP1L1,

- FLNA, FLNB, FLNC,

- GAS2, GAS2L1, GAS2L2, GAS2L3,

- IQGAP1, IQGAP2, IQGAP3,

- LCP1, LIMCH1, LMO7, LRCH1, LRCH2, LRCH3, LRCH4,

- MACF1, MAPRE1, MAPRE2, MAPRE3, MICAL1, MICAL2, MICAL2PV1, MICAL2PV2, MICAL3, MICALL1, MICALL2,

- NAV2, NAV3,

- PARVA, PARVB, PARVG, PLEC1, PLS1, PLS3, PP14183,

- SMTN, SMTNL2, SPECC1, SPECC1L, SPNB4, SPTB, SPTBN1, SPTBN2, SPTBN4, SPTBN5, SYNE1, SYNE2,

- TAGLN, TAGLN2, TAGLN3,

- UTRN, and

- VAV1, VAV2, VAV3

PDZ Domain

The PDZ domain is a common structural domain of 80-90 amino-acids found in the signaling proteins of bacteria, yeast, plants, viruses and animals. Proteins containing PDZ domains play a key role in anchoring receptor proteins in the membrane to cytoskeletal components. Proteins with these domains help hold together and organize signaling complexes at cellular membranes. These domains play a key role in the formation and function of signal transduction complexes. PDZ domains also play a highly significant role in the anchoring of cell surface receptors (such as Cftr and FZD7) to the actin cytoskeleton via mediators like NHERF and ezrin.

PDZ is an initialism combining the first letters of the first three proteins discovered to share the domain — post synaptic density protein (PSD95), Drosophila disc large tumor suppressor (Dlg1), and zonula occludens-1 protein (zo-1). PDZ domains have previously been referred to as DHR (Dlg homologous region) or GLGF (glycine-leucine-glycine-phenylalanine) domains.

In general PDZ domains bind to a short region of the C-terminus of other specific proteins. These short regions bind to the PDZ domain by beta sheet augmentation. This means that the beta sheet in the PDZ domain is extended by the addition of a further beta strand from the tail of the binding partner protein. The C-terminal carboxylate group is bound by a nest (protein structural motif) in the PDZ domain.

PDZ is an acronym derived from the names of the first proteins in which the domain was observed. Post-synaptic density protein 95 (PSD-95) is a synaptic protein found only in the brain. Drosophila disc large tumor suppressor (Dlg1) and zona occludens 1 (ZO-1) both play an important role at cell junctions and in cell signaling complexes. Since the discovery of PDZ domains more than 20 years ago, researchers have successfully identified hundreds of PDZ domains. The first published use of the phrase "PDZ domain" was not in a paper, but a letter. In September 1995, Dr. Mary B. Kennedy of the California Institute of Technology wrote a letter of correction to Trends in Biomedical Sciences. Earlier that year, another set of scientists had claimed to discover a new protein domain which they called a DHR domain. Dr. Kennedy refuted that her lab had previously described exactly the same domain as a series of "GLGF repeats". She continued to explain that in order to "better reflect the origin and distribution of the domain," the new title of the domain would be changed. Thus, the name "PDZ domain" was introduced to the world.

Structure

PDZ domain structure is partially conserved across the various proteins that contain them. They usually have 4 β-strands and one short and one long α-helix. Apart from this conserved fold, the secondary structure differs across PDZ domains. This domain tends to be globular with a diameter of about 35 Å.

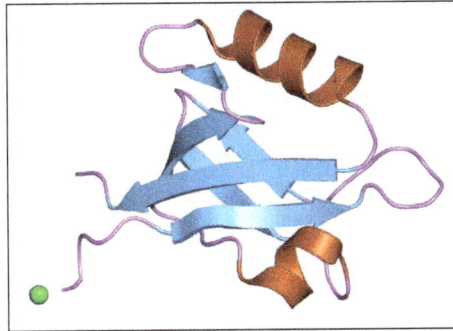

6 β-strands (blue) and two α-helix (red) are
the common motif for PDZ domains.

When studied, PDZ domains are usually isolated as monomers, however some PDZ proteins form dimers. The function of PDZ dimers as compared to monomers is not yet known.

A commonly accepted theory for the binding pocket of the PDZ domain is that it is constituted by several hydrophobic amino acids, apart from the GLGF sequence mentioned earlier, the mainchain atoms of which form a nest (protein structural motif) binding the C-terminal carboxylate of the protein or peptide ligand. Most PDZ domains have such a binding site located between one of the β-strands and the long α-helix.

Functions

PDZ domains have two main functions: Localizing cellular elements, and regulating cellular pathways.

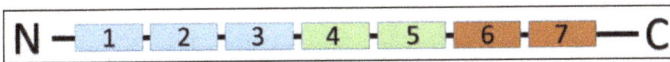

An example of a protein (GRIP) with seven PDZ domains.

The first discovered function of the PDZ domains was to anchor receptor proteins in the membrane to cytoskeletal components. PDZ domains also have regulatory functions on different signaling pathways. Any protein may have one or several PDZ domains, which can be identical or unique .This variety allows these proteins to be very versatile in their interactions. Different PDZ domains in the same protein can have different roles, each binding a different part of the target protein or a different protein altogether.

Localization

PDZ domains play a vital role in organizing and maintaining complex scaffolding formations.

PDZ domains are found in diverse proteins, but all assist in localization of cellular elements. PDZ domains are primarily involved in anchoring receptor proteins to the cytoskeleton. For cells to function properly it is important for components—proteins and other molecules— to be in the right place at the right time. Proteins with PDZ domains bind different components to ensure correct arrangements. In the neuron, making sense of neurotransmitter activity requires specific receptors to be located in the lipid membrane at the synapse. PDZ domains are crucial to this receptor localization process. Proteins with PDZ domains generally associate with both the C-terminus of the receptor and cytoskeletal elements in order to anchor the receptor to the cytoskeleton and keep it in place. Without such an interaction, receptors would diffuse out of the synapse due to the fluid nature of the lipid membrane.

PDZ domains are also utilized to localize elements other than receptor proteins. In the human brain, nitric oxide often acts in the synapse to modify cGMP levels in response to NMDA receptor activation. In order to ensure a favorable spatial arrangements, neuronal nitric oxide synthase (nNOS) is brought close to NMDA receptors via interactions with PDZ domains on PSD-95, which concurrently binds nNOS and NMDA receptors. With nNOS located closely to NMDA receptors, it will be activated immediately after calcium ions begin entering the cell.

Regulation

PDZ domains are directly involved in the regulation of different cellular pathways. This mechanism of this regulation varies as PDZ domains are able to interact with a range of cellular components. This regulation is usually a result of the co-localization of multiple signaling molecules such as in the example with nNos and NMDA receptors. Some examples of signaling pathway regulation executed by the PDZ domain include phosphatase enzyme activity, mecahnosensory signaling, and the sorting pathway of endocytosed receptor proteins.

The signaling pathway of the human protein tyrosine phosphatase non-receptor type 4 (PTPN4) is regulated by PDZ domains. This protein is involved in regulating cell death. Normally the PDZ domain of this enzyme is unbound. In this unbound state the enzyme is active and prevents cell signaling for apoptosis. Binding the PDZ domain of this phosphatase results in a loss of enzyme activity, which leads to apoptosis. The normal regulation of this enzyme prevents cells from prematurely going through apoptosis. When the regulation of the PTPN4 enzyme is lost, there is increased oncogenic activity as the cells are able to proliferate.

PDZ domains also have a regulatory role in mechanosensory signaling in proprioceptors and vestibular and auditory hair cells. The protein Whirlin (WHRN) localizes in the post-synaptic neurons of hair cells that transform mechanical movement into action potentials that the body can interpret. WHRN proteins contains three PDZ domains. The domains located near the N-terminus bind to receptor proteins and other signaling components. When the one of these PDZ domains is inhibited, the signaling pathways of the neurons are disrupted, resulting in auditory, visual, and vestibular impairment. This regulation is thought to be based on the physical positioning WHRN and the selectivity of its PDZ domain.

Regulation of receptor proteins occurs when the PDZ domain on the EBP50 protein binds to the C-terminus of the beta-2 adrenergic receptor (ß2-AR). EBP50 also associates with a complex that connects to actin, thus serving as a link between the cytoskeleton and ß2-AR. The ß2-AR receptor is eventually endocytosed, where it will either be consigned to a lysosome for degradation or recycled back to the cell membrane. Scientists have demonstrated that when the Ser-411 residue of the ß2-AR PDZ binding domain, which interacts directly with EBP50, is phosphorylated, the receptor is degraded. If Ser-411 is left unmodified, the receptor is recycled. The role played by PDZ domains and their binding sites indicate a regulative relevance beyond simply receptor protein localization.

Regulation of PDZ Domain Activity

PDZ domain function can be both inhibited and activated by various mechanisms. Two of the most prevalent include allosteric interactions and posttraslational modifications.

Post-translational Modifications

The most common post-traslational modification seen on PDZ domains is phosphorylation. This modification is primarily an inhibitor of PDZ domain and ligand activity. In some examples, phosphorylation of amino acid side chains eliminates the ability of the PDZ domain to form hydrogen bonds, disrupting the normal binding patterns. The end result is a loss of PDZ domain function and further signaling. Another way phosphorylation can disrupt regular PDZ domain funcition is by altering the charge ratio and further affecting binding and signaling. In rare cases researchers have seen post-translational modifications activate PDZ domain activity but these cases are few.

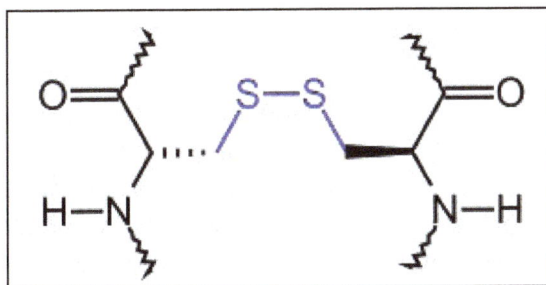

Disulfide bridges inhibit PDZ domain function.

Another post-translational modification that can regulate PDZ domains is the formation of disulfide bridges. Many PDZ domains contain cysteines and are susceptible to disulfide bond formation in oxidizing conditions. This modification acts primarily as an inhibitor of PDZ domain function.

Allosteric Interactions

Protein-protein interactions have been observed to alter the effectiveness of PDZ domains binding to ligands. These studies show that allosteric effects of certain proteins can affect the binding affinity for different substrates. Different PDZ domains can even have this allosteric effect on each other. This PDZ-PDZ interaction only acts as an inhibitor. Other experiments have shown that certain enzymes can enhance the binding of PDZ domains. Researchers found that the protein ezrin enhances the binding of the PDZ protein NHERF1.

PDZ Proteins

PDZ proteins are a family of proteins that contain the PDZ domain. This sequence of amino-acids is found in many thousands of known proteins. PDZ domain proteins are widespread in eukaryotes and eubacteria, whereas there are very few examples of the protein in archaea. PDZ domains are often associated with other protein domains and these combinations allow them to carry out their specific functions. Three of the most well documented PDZ proteins are PSD-95, GRIP, and HOMER.

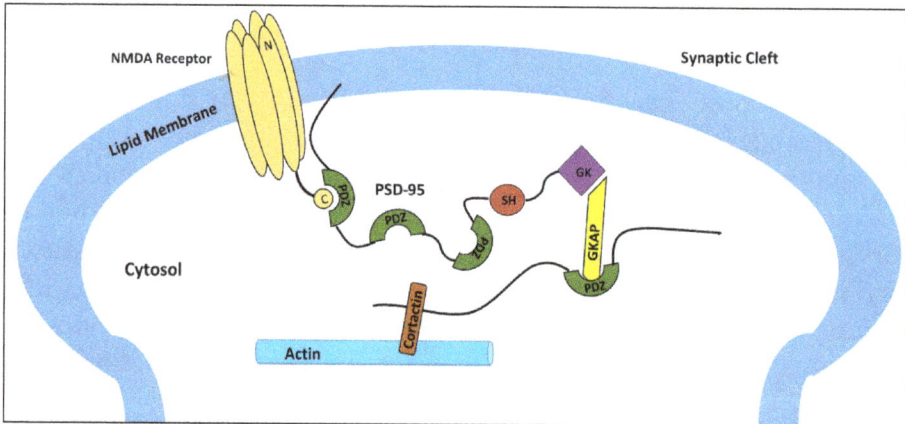

Basic functioning of PSD-95 in forming a complex between NMDA Receptor and Actin.

PSD-95 is a brain synaptic protein with three PDZ domains, each with unique properties and structures that allow PSD-95 to function in many ways. In general, the first two PDZ domains interact with receptors and the third interacts with cytoskeleton-related proteins. The main receptors associated with PSD-95 are NMDA receptors. The first two PDZ domains of PSD-95 bind to the C-terminus of NMDA receptors and anchor them in the membrane at the point of neurotransmitter release. The first two PDZ domains can also interact in a similar fashion with Shaker-type K+ channels. A PDZ interaction between PSD-95, nNOS and syntrophin is mediated by the second PDZ domain. The third and final PDZ domain links to cysteine-rich PDZ-binding protein (CRIPT), which allows PSD-95 to associate with the cytoskeleton.

Glutamate receptor interacting protein (GRIP) is a post-synaptic protein with that interacts with AMPA receptors in a fashion analogous to PSD-95 interactions with NMDA receptors. When researchers noticed apparent structural homology between the C-termini of AMPA receptors and NMDA receptors, they attempted to determine if a similar PDZ interaction was occurring. A yeast two-hybrid system helped them discover that out of GRIP's seven PDZ domains, two (domains four and five) were essential for binding of GRIP to the AMPA subunit called GluR2. This interaction is vital for proper localization of AMPA receptors, which play a large part in memory storage. Other researchers discovered that domains six and seven of GRIP are responsible for connecting GRIP to a family of receptor tyrosine kinases called ephrin receptors, which are important signaling proteins. A clinical study concluded that Fraser syndrome, an autosomal recessive syndrome that can cause severe deformations, can be caused by a simple mutation in GRIP.

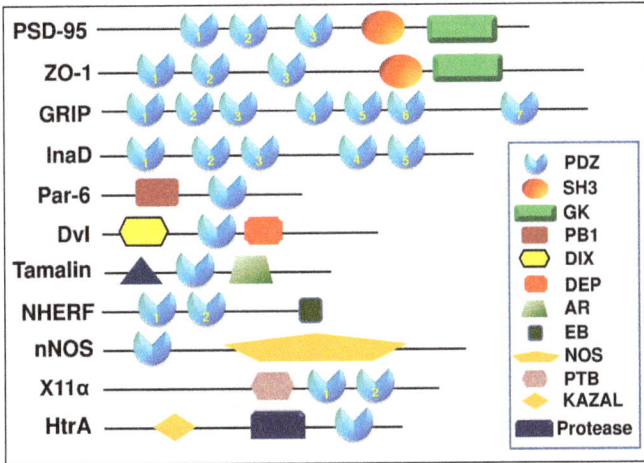

Examples of PDZ domain-containing proteins Proteins are indicated by black lines scaled to the length of the primary sequence of the protein. Different shapes refer to different protein domains.

HOMER differs significantly from many known PDZ proteins, including GRIP and PSD-95. Instead of mediating receptors near ion channels, as is the case with GRIP and PSD-95, HOMER is involved in metabotropic glutamate signaling. Another unique aspect of HOMER is that it only contains a single PDZ domain, which mediates interactions between HOMER and type 5 metabotropic glutamate receptor (mGluR5). The single GLGF repeat on HOMER binds amino acids on the C-terminus of mGluR5. HOMER expression is measured at high levels during embryologic stages in rats, suggesting an important developmental function.

Human PDZ Proteins

There are roughly 260 PDZ domains in humans. Several proteins contain multiple PDZ domains, so the number of unique PDZ-containing proteins is closer to 180. In the table below are some of the better studied members of this family:

Studied PDZ Proteins			
Erbin	GRIP	Htra1	Htra2
Htra3	PSD-95	SAP97	CARD10
CARD11	CARD14	PTP-BL	

The table below contains all known PDZ proteins in humans (alphabetical):

PDZ Proteins in Humans							
AAG12	AHNAK	AHNAK2	AIP1	ALP	APBA1	APBA2	APBA3
CASK	CLP-36	CNKSR2	CNKSR3	CRTAM	DFNB31	DLG1	DLG2
ERBB2IP	FRMPD1	FRMPD2	FRMPD2L1	FRMPD3	FRMPD4	GIPC1	GIPC2
HTRA3	HTRA4	IL16	INADL	KIAA1849	LDB3	LIMK1	LIMK2
MAGI1	MAGI2	MAGI3	MAGIX	MAST1	MAST2	MAST3	MAST4
MPP5	MPP6	MPP7	MYO18A	NHERF1	NOS1	PARD3	PARD6A
PDLIM7	PDZD11	PDZD2	PDZD3	PDZD4	PDZD5A	PDZD7	PDZD8
PRX	PSCDBP	PTPN13	PTPN3	PTPN4	RAPGEF2	RGS12	RGS3
SDCBP2	SHANK1	SHANK2	SHANK3	SHROOM2	SHROOM3	SHROOM4	SIPA1
SNTB2	SNTG1	SNTG2	SNX27	SPAL2	STXBP4	SYNJ2BP	SYNPO2
TRPC4	TRPC5	USH1C	WHRN				

PDZ Proteins in Humans							
AAG12	ARHGAP21	ARHGAP23	ARHGEF11	ARHGEF12	CARD10	CARD11	CARD14
CASK	DLG3	DLG4	DLG5	DVL1	DVL1L1	DVL2	DVL3
ERBB2IP	GIPC3	GOPC	GRASP	GRIP1	GRIP2	HTRA1	HTRA2
HTRA3	LIN7A	LIN7B	LIN7C	LMO7	LNX1	LNX2	LRRC7
MAGI1	MCSP	MLLT4	MPDZ	MPP1	MPP2	MPP3	MPP4
MPP5	PARD6B	PARD6G	PDLIM1	PDLIM2	PDLIM3	PDLIM4	PDLIM5
PDLIM7	PDZK1	PDZRN3	PDZRN4	PICK1	PPP1R9A	PPP1R9B	PREX1
PRX	RHPN1	RIL	RIMS1	RIMS2	SCN5A	SCRIB	SDCBP
SDCBP2	SIPA1L1	SIPA1L2	SIPA1L3	SLC9A3R1	SLC9A3R2	SNTA1	SNTB1
SNTB2	SYNPO2L	TAX1BP3	TIAM1	TIAM2	TJP1	TJP2	TJP3
TRPC4							

There is currently one known virus of PDZ domains.

LIM Domain

LIM domains are protein structural domains, composed of two contiguous zinc finger domains, separated by a two-amino acid residue hydrophobic linker. They are named after their initial discovery in the proteins Lin11, Isl-1 & Mec-3. LIM-domain containing proteins have been shown to play roles in cytoskeletal organisation, organ development and oncogenesis. LIM-domains mediate protein–protein interactions that are critical to cellular processes.

LIM domains have highly divergent sequences, apart from certain key residues. The sequence divergence allow a great many different binding sites to be grafted onto the same basic domain. The conserved residues are those involved in zinc binding or the hydrophobic core of the protein. The sequence signature of LIM domains is as follows:

$$[C]\text{-}[X]_{2-4}\text{-}[C]\text{-}[X]_{13-19}\text{-}[W]\text{-}[H]\text{-}[X]_{2-4}\text{-}[C]\text{-}[F]\text{-}[LVI]\text{-}[C]\text{-}[X]_{2-4}\text{-}[C]\text{-}[X]_{13-20}\text{-}C\text{-}[X]_{2-4}\text{-}[C]$$

LIM domains frequently occur in multiples, as seen in proteins such as TES, LMO4, and can also be attached to other domains in order to confer a binding or targeting function upon them, such as LIM-kinase.

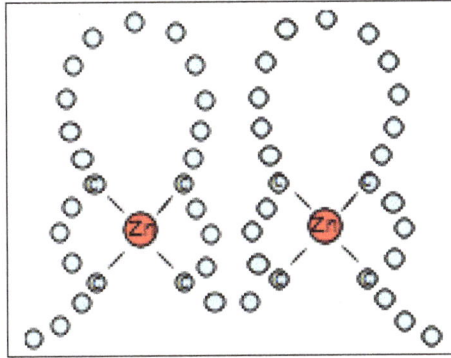

LIM domain organisation.

The LIM superclass of genes have been classified into 14 classes: ABLIM, CRP, ENIG-MA, EPLIN, LASP, LHX, LMO, LIMK, LMO7, MICAL, PXN, PINCH, TES, and ZYX. Six of these classes (i.e., ABLIM, MICAL, ENIGMA, ZYX, LHX, LMO7) originated in the stem lineage of animals, and this expansion is thought to have made a major contribution to the origin of animal multicellularity.

LIM domains are also found in various bacterial lineages where they are typically fused to a metallopeptidase domain. Some versions show fusions to an inactive P-loop NT-Pase at their N-terminus and a single transmembrane helix. These domain fusions suggest that the prokaryotic LIM domains are likely to regulate protein processing at the cell membrane. The domain architectural syntax is remarkable parallel to those of the prokaryotic versions of the B-box zinc finger and the AN1 zinc finger domains.

SH2 Domain

The SH2 (Src Homology 2) domain is a structurally conserved protein domain contained within the Src oncoprotein and in many other intracellular signal-transducing proteins. SH2 domains allow proteins containing those domains to dock to phosphorylated tyrosine residues on other proteins. SH2 domains are commonly found in adaptor proteins that aid in the signal transduction of receptor tyrosine kinase pathways.

SH2 is conserved by signalization of protein tyrosine kinase, which are binding on phosphotyrosine (pTyr). In the human proteome the class of pTyr-selective recognition domains is represented by SH2 domains. The N-terminal SH2 domains of cytoplasmic tyrosine kinase was at the beginning of evolution evolved with the occurrence of tyrosine phosphorylation. At the beginning it was supposed that, these domains serve as a substrate for their target kinase.

Protein-protein interactions play a major role in cellular growth and development. Modular domains, which are the subunits of a protein, moderate these protein interactions by identifying short peptide sequences. These peptide sequences determine the binding partners of each protein. One of the more prominent domains is the SH2 domain. SH2 domains play a vital role in cellular communication. Its length is approximately 100 amino acids long and it is found within 111 human proteins. Regarding its structure, it contains 2 alpha helices and 7 beta strands. Research has shown that it has a high affinity to phosphorylated tyrosine residues and it is known to identify a sequence of 3-6 amino acids within a peptide motif.

Binding and Phosphorylation

SH2 domains typically bind a phosphorylated tyrosine residue in the context of a longer peptide motif within a target protein, and SH2 domains represent the largest class of known pTyr-recognition domains.

Phosphorylation of tyrosine residues in a protein occurs during signal transduction and is carried out by tyrosine kinases. In this way, phosphorylation of a substrate by tyrosine kinases acts as a switch to trigger binding to an SH2 domain-containing protein. Many tyrosine containing short linear motifs that bind to SH2 domains are conserved across a wide variety of higher Eukaryotes. The intimate relationship between tyrosine kinases and SH2 domains is supported by their coordinate emergence during eukaryotic evolution.

Diversity

SH2 domains are not present in yeast and appear at the boundary between protozoa and animalia in organisms such as the social amoeba Dictyostelium discoideum.

A detailed bioinformatic examination of SH2 domains of human and mouse reveals 120 SH2 domains contained within 115 proteins encoded by the human genome, representing a rapid rate of evolutionary expansion among the SH2 domains.

A large number of SH2 domain structures have been solved and many SH2 proteins have been knocked out in mice.

Function

The function of SH2 domains is to specifically recognize the phosphorylated state of tyrosine residues, thereby allowing SH2 domain-containing proteins to localize to tyrosine-phosphorylated sites. This process constitutes the fundamental event of signal transduction through a membrane, in which a signal in the extracellular compartment is "sensed" by a receptor and is converted in the intracellular compartment to a different chemical form, i.e. that of a phosphorylated tyrosine. Tyrosine phosphorylation leads to activation of a cascade of protein-protein interactions whereby SH2 domain-containing proteins are recruited to tyrosine-phosphorylated sites. This process initiates a series of events which eventually result in altered patterns of gene expression or other cellular responses. The SH2 domain, which was first identified in the oncoproteins Src and Fps, is about 100 amino-acid residues long. It functions as a regulatory module of intracellular signaling cascades by interacting with high affinity to phosphotyrosine-containing target peptides in a sequence-specific and strictly phosphorylation-dependent manner.

Applications

SH2 domains, and other binding domains, have been used in protein engineering to create protein assemblies. Protein assemblies are formed when several proteins bind to one another to create a larger structure (called a supramolecular assembly). Using molecular biology techniques, fusion proteins of specific enzymes and SH2 domains have been created, which can bind to each other to form protein assemblies.

Since SH2 domains require phosphorylation in order for binding to occur, the use of kinase and phosphatase enzymes gives researchers control over whether protein assemblies will form or not. High affinity engineered SH2 domains have been developed and utilized for protein assembly applications.

The goal of most protein assembly formation is to increase the efficiency of metabolic pathways via enzymatic co-localization. Other applications of SH2 domain mediated protein assemblies have been in the formation of high density fractal-like structures, which have extensive molecular trapping properties.

Examples:

Human proteins containing this domain include:

- ABL1; ABL2

- BCAR3; BLK; BLNK; BMX; BTK

- CHN2; CISH; CRK; CRKL; CSK

- DAPP1

- FER; FES; FGR; FRK; FYN

- GRAP; GRAP2; GRB10; GRB14; GRB2; GRB7

- HCK; HSH2D

- INPP5D; INPPL1; ITK; JAK2; LCK; LCP2; LYN

- MATK; NCK1; NCK2

- PIK3R1; PIK3R2; PIK3R3; PLCG1; PLCG2; PTK6; PTPN11; PTPN6; RASA1

- SH2B1; SH2B2; SH2B3; SH2D1A; SH2D1B; SH2D2A; SH2D3A; SH2D3C; SH2D4A; SH2D4B; SH2D5; SH2D6; SH3BP2; SHB; SHC1; SHC3; SHC4; SHD; SHE

- SLA; SLA2

- SOCS1; SOCS2; SOCS3; SOCS4; SOCS5; SOCS6; SOCS7

- SRC; SRMS

- STAT1; STAT2; STAT3; STAT4; STAT5A; STAT5B; STAT6

- SUPT6H; SYK

- TEC; TENC1; TNS; TNS1; TNS3; TNS4; TXK

- VAV1; VAV2; VAV3

- YES1; ZAP70

FERM Domain

Previously known as the B4.1 (band 4.1) homology and ERM domain, the FERM domain is named for the four proteins from which this domain was originally described: Band 4.1 (F), Ezrin (E), Radixin (R), and Moesin (M). The FERM domain is approximately 150 amino acids in length and is found in a number of cytoskeletal-associated proteins that are localized to the plasma membrane and cytoskeleton interface. The FERM domain is responsible for the PIP2 regulated binding of ERM (Ezrin/Radixin/Moesin) proteins to the membrane, which associates the cytoskeleton with the membrane by linking actin filaments to adhesion proteins. The FERM domain containing adaptor protein Merlin has been shown to act as a tumour suppressor. FERM domains contain three lobes, with the N-terminal lobe resembling ubiquitin, the central lobe resembling acyl-CoA binding proteins and the C-terminal lobe having a structure similar to that of PTB domains. The structure of the Radixin FERM domain bound to IP3 has been solved; phosphoinositide binding is not mediated by the PH-fold subdomain of FERM but occurs at a cleft between two subdomains on a relatively flat face of the module. The FERM domain is also thought to bind to adhesion proteins in a PIP2-regulated fashion to provide a link between cytoskeletal signals and membrane dynamics.

Domain Proteins

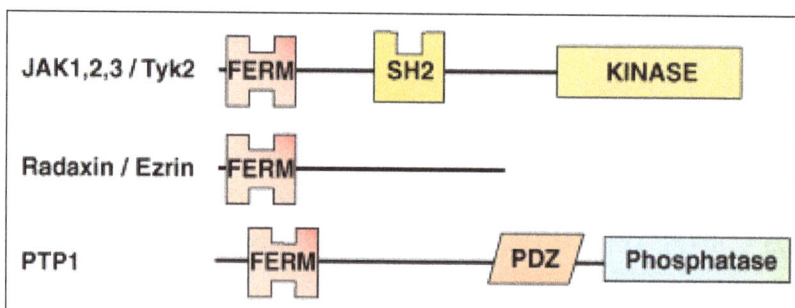

Binding

FERM Domain Proteins	Motif	Binding Partners
Talin	NPxY	integrin β subunits2.
Radixin		IP_3, $PI(4,5)P_2$
ERM		$PI(4,5)P_2$
Willin		$PI(3)P$, $PI(4)P$, $PI(5)P$
Moesin		$PI(3)P$, $PI(4)P$, $PI(5)P$
PTLP1		$PI(4,5)P_2$
Radixin		$PI(1,4,5)P_3$

PROTEIN DYNAMICS

Proteins are generally thought to adopt unique structures determined by their amino acid sequences, as outlined by Anfinsen's dogma. However, proteins are not strictly static objects, but rather populate ensembles of (sometimes similar) conformations. Transitions between these states occur on a variety of length scales (tenths of Å to nm) and time scales (ns to s), and have been linked to functionally relevant phenomena such as allosteric signaling and enzyme catalysis.

The study of protein dynamics is most directly concerned with the transitions between these states, but can also involve the nature and equilibrium populations of the states themselves. These two perspectives—kinetics and thermodynamics, respectively—can be conceptually synthesized in an "energy landscape" paradigm: highly populated states and the kinetics of transitions between them can be described by the depths of energy wells and the heights of energy barriers, respectively.

Local Flexibility: Atoms and Residues

Portions of protein structures often deviate from the equilibrium state. Some such excursions are harmonic, such as stochastic fluctuations of chemical bonds and bond angles. Others are anharmonic, such as sidechains that jump between separate discrete energy minima, or rotamers.

Evidence for local flexibility is often obtained from NMR spectroscopy. Flexible and potentially disordered regions of a protein can be detected using the random coil index. Flexibility in folded proteins can be identified by analyzing the spin relaxation of individual atoms in the protein. Flexibility can also be observed in very high-resolution electron density maps produced by X-ray crystallography, particularly when diffraction data is collected at room temperature instead of the traditional cryogenic temperature (typically near 100 K). Information on the frequency distribution and dynamics of local protein flexibility can be obtained using Raman and optical Kerr-effect spectroscopy in the terahertz frequency domain.

Regional Flexibility: Intra-domain Multi-residue Coupling

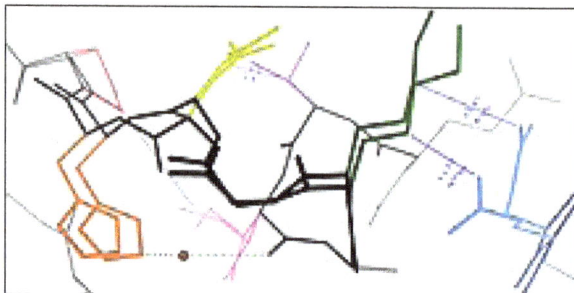

A network of alternative conformations in catalase (Protein Data Bank code: 1gwe) with diverse properties. Multiple phenomena define the network: van der Waals interactions (blue dots and line segments) between sidechains, a hydrogen bond (dotted green line) through a partial-occupancy water (brown), coupling through the locally mobile backbone (black), and perhaps electrostatic forces between the Lys (green) and nearby polar residues (blue: Glu, yellow: Asp, purple: Ser). This particular network is distal from the active site and is therefore putatively not critical for function.

Many residues are in close spatial proximity in protein structures. This is true for most residues that are contiguous in the primary sequence, but also for many that are distal in sequence yet are brought into contact in the final folded structure. Because of this proximity, these residues's energy landscapes become coupled based on various biophysical phenomena such as hydrogen bonds, ionic bonds, and van der Waals interactions. Transitions between states for such sets of residues therefore become correlated.

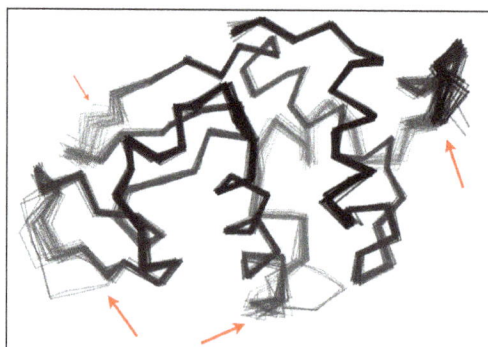

An "ensemble" of 44 crystal structures of hen egg white lysozyme from the Protein Data Bank, showing that different crystallization conditions lead to different conformations for various surface-exposed loops and termini (red arrows).

This is perhaps most obvious for surface-exposed loops, which often shift collectively to adopt different conformations in different crystal structures. However, coupled conformational heterogeneity is also sometimes evident in secondary structure. For example, consecutive residues and residues offset by 4 in the primary sequence often interact in α helices. Also, residues offset by 2 in the primary sequence point their side-chains toward the same face of β sheets and are close enough to interact sterically, as are residues on adjacent strands of the same β sheet.

When these coupled residues form pathways linking functionally important parts of a protein, they may participate in allosteric signaling. For example, when a molecule of oxygen binds to one subunit of the hemoglobin tetramer, that information is allosterically propagated to the other three subunits, thereby enhancing their affinity for oxygen. In this case, the coupled flexibility in hemoglobin allows for cooperative oxygen binding, which is physiologically useful because it allows rapid oxygen loading in lung tissue and rapid oxygen unloading in oxygen-deprived tissues (e.g. muscle).

Global Flexibility: Multiple Domains

The presence of multiple domains in proteins gives rise to a great deal of flexibility and mobility, leading to protein domain dynamics. Domain motions can be inferred by comparing different structures of a protein (as in Database of Molecular Motions), or they can be directly observed using spectra measured by neutron spin echo spectroscopy. They can also be suggested by sampling in extensive molecular dynamics trajectories and principal component analysis. Domain motions are important for:

- ABC transporters.
- Catalysis.
- Cellular locomotion and motor proteins.
- Formation of protein complexes.
- Ion channels.
- Mechanoreceptors and mechanotransduction.
- Regulatory activity.
- transport of metabolites across cell membranes.

One of the largest observed domain motions is the 'swivelling' mechanism in pyruvate phosphate dikinase. The phosphoinositide domain swivels between two states in order to bring a phosphate group from the active site of the nucleotide binding domain to that of the phosphoenolpyruvate/pyruvate domain. The phosphate group is moved over a distance of 45 Å involving a domain motion of about 100 degrees around a single residue. In enzymes, the closure of one domain onto another captures a substrate by an induced fit, allowing the reaction to take place in a controlled way. A detailed analysis by Gerstein led to the classification of two basic types of domain motion; hinge and shear. Only a relatively small portion of the chain, namely the inter-domain linker and side chains undergo significant conformational changes upon domain rearrangement.

Hinges by Secondary Structures

A study by Hayward found that the termini of α-helices and β-sheets form hinges in a large number of cases. Many hinges were found to involve two secondary structure elements acting like hinges of a door, allowing an opening and closing motion to occur. This can arise when two neighbouring strands within a β-sheet situated in one domain, diverge apart as they join the other domain. The two resulting termini then form the bending regions between the two domains. α-helices that preserve their hydrogen bonding network when bent are found to behave as mechanical hinges, storing 'elastic energy' that drives the closure of domains for rapid capture of a substrate.

Helical to Extended Conformation

The interconversion of helical and extended conformations at the site of a domain boundary is not uncommon. In calmodulin, torsion angles change for five residues in the middle of a domain linking α-helix. The helix is split into two, almost perpendicular, smaller helices separated by four residues of an extended strand.

Shear Motions

Shear motions involve a small sliding movement of domain interfaces, controlled by the amino acid side chains within the interface. Proteins displaying shear motions often have a layered architecture: stacking of secondary structures. The interdomain linker has merely the role of keeping the domains in close proximity.

Domain Motion and Functional Dynamics in Enzymes

The analysis of the internal dynamics of structurally different, but functionally similar enzymes has highlighted a common relationship between the positioning of the active site and the two principal protein sub-domains. In fact, for several members of the hydrolase superfamily, the catalytic site is located close to the interface separating the two principal quasi-rigid domains. Such positioning appears instrumental for maintaining the precise geometry of the active site, while allowing for an appreciable functionally oriented modulation of the flanking regions resulting from the relative motion of the two sub-domains.

Implications for Macromolecular Evolution

Evidence suggests that protein dynamics are important for function, e.g. enzyme catalysis in DHFR, yet they are also posited to facilitate the acquisition of new functions by molecular evolution. This argument suggests that proteins have evolved to have stable, mostly unique folded structures, but the unavoidable residual flexibility leads to some degree of functional promiscuity, which can be amplified/harnessed/diverted by subsequent mutations.

However, there is growing awareness that intrinsically unstructured proteins are quite prevalent in eukaryotic genomes, casting further doubt on the simplest interpretation of Anfinsen's dogma: "sequence determines structure (singular)". In effect, the new paradigm is characterized by the addition of two caveats: "sequence and cellular environment determine structural ensemble".

GST-TAGGED PROTEINS

Protein purification with affinity tags such as glutathione S-transferase (GST), histidine (HIS), and other affinity tags, enables purification of proteins with both known and unknown biochemical properties. Therefore, this methodology has become a widely used research tool for determining the biological function of uncharacterized proteins. GST is a 211 amino acid protein (26 kDa) whose DNA sequence is frequently integrated into expression vectors for production of recombinant proteins. The result of expression from this vector is a GST-tagged fusion protein in which the functional GST protein (26 kDa) is fused to the N-terminus of the recombinant protein.

Because GST folds rapidly into a stable and highly soluble protein upon translation, inclusion of the GST tag often promotes greater expression and solubility of recombinant proteins than expression without the tag. In addition, GST-tagged fusion proteins can be purified or detected based on the ability of GST (an enzyme) to bind its substrate, glutathione (GSH).

GST-fusion Protein Purification

Glutathione is a tripeptide (Glu-Cys-Gly) that is the specific substrate for glutathione S-transferase (GST). When reduced glutathione is immobilized through its sulfhydryl group to a solid support, such as cross-linked beaded agarose, it can be used to capture pure GST or GST-tagged proteins via the enzyme-substrate binding reaction.

Binding is most effective in near-neutral buffers (physiologic conditions) such as Tris-buffered saline (TBS) pH 7.5. Because binding depends on preserving the essential structure and enzymatic function of GST, protein denaturants are not compatible.

After washing an affinity column to remove non-bound sample components, the purified GST-fusion protein can be dissociated and recovered (eluted) from a glutathione column by addition of excess reduced glutathione. The free glutathione competitively displaces the immobilized glutathione binding interaction with GST, allowing the fusion protein to emerge from the affinity column.

Immobilized Glutathione. Cross-linked beaded
agarose bound to reduced glutathione.

This affinity system commonly yields greater than 90% pure GST-tagged recombinant protein from crude bacterial or mammalian cell lysate samples. Glutathione-based affinity purification of GST-tagged fusion proteins is easily done at either small, medium or large scales to produce microgram, milligram or gram quantities.

At 26 kDa, GST is considerably larger than many other fusion protein affinity tags. For reasons that have not been fully characterized in the literature, the structure of the GST-fusion tag often degrades upon denaturation and reduction for protein gel electrophoresis (e.g., SDS-PAGE). As a result, electrophoresed samples of GST-fusion proteins often appear as a ladder of lower MW bands below the full-sized fusion protein.

When the GST tag is not required or desired as part of the recombinant protein after purification, it can be removed if a cleavage site for a specific protease is included between the protein and GST tag in the design of the DNA vector. For example, HRV 3C protease specifically cleaves the sequence Leu-Glu-Val-Leu-Phe-Gln-↓-Gly-Pro

Vendor	Thermo Scientific Pierce	GE Healthcare	Qiagen	Clontech	Sigma
Yield	537µg	562µg	285µg	299µg	410µg
Purity	93%	93%	90%	91%	94%

Glutathione agarose delivers high yield and high purity GST-fusion proteins. *E. coli* lysate (14.4 mg total protein) containing overexpressed GST was incubated with 50 µL

GSH resin from various suppliers and purified per manufacturers' instructions. The amount of GST eluted from the resin (yield) was quantified by Thermo Scientific Coomassie Plus Protein Assay. Purity was assessed by densitometry of the stained gel lanes. M= MW marker; L=Lysate load; FT=Flow-through; E=Elution.

Other GST-tagged Protein Techniques

Besides affinity purification, other applications for GST-tagged fusion proteins are made possible with the aid of glutathione-ligand chemistries or GST-tag-specific antibodies:

- Microplate coating: Glutathione-coated or anti-GST antibody microplates may be used for determining the presence and concentration of GST or GST-fusion proteins from cell lysates. Because the plates are pre-blocked, initial purification of the cell lysates is not necessary. Also, using these plates, which are available in both 8-well and 96-well formats, to immobilize GST-fusion proteins is useful for screening sera for antibodies to the fusion protein.

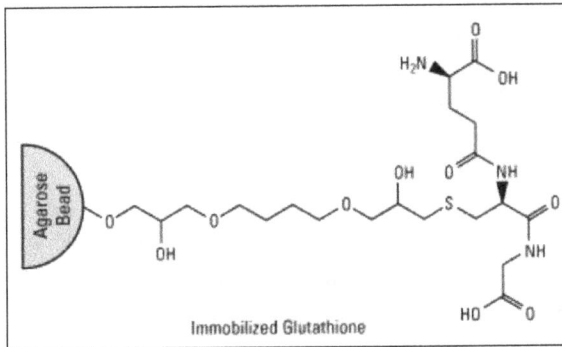

Immobilized Glutathione

Binding of GST to glutathione-coated plates. Thermo Scientific Pierce GST microplates enable fusion proteins to be captured from crude or semi-purified samples for plate and reporter assays of various kinds.

- Protein interaction pull-down: Specific GST-tagged proteins and glutathione agarose resin are the basis of kits designed to purify, identify and measure specific protein interaction complexes.

- ELISA or western blot detection: Anti-GST antibodies are available for these applications.

HIS-TAGGED PROTEINS

The DNA sequence specifying a string of six to nine histidine residues is frequently used in vectors for production of recombinant proteins. The result is expression of a recombinant protein with a 6xHis or poly-His-tag fused to its N- or C-terminus.

Expressed His-tagged proteins can be purified and detected easily because the string of histidine residues binds to several types of immobilized metal ions, including nickel, cobalt and copper, under specific buffer conditions. In addition, anti-His-tag antibodies are commercially available for use in assay methods involving His-tagged proteins. In either case, the tag provides a means of specifically purifying or detecting the recombinant protein without a protein-specific antibody or probe.

Immobilized Metal Affinity Chromatography

Supports such as beaded agarose or magnetic particles can be derivatized with chelating groups to immobilize the desired metal ions, which then function as ligands for binding and purification of biomolecules of interest. This basis for affinity purification is known as immobilized metal affinity chromatography (IMAC). IMAC is a widely-used method for rapidly purifying polyhistidine affinity-tagged proteins, resulting in 100-fold enrichments in a single purification step.

The chelators most commonly used as ligands for IMAC are nitrilotriacetic acid (NTA) and iminodiacetic acid (IDA). Once IDA-agarose or NTA-agarose resin is prepared, it can be "loaded" with the desired divalent metal (e.g., Ni, Co, Cu, and Fe). Using nickel as the example metal, the resulting affinity support is usually called Ni-chelate, Ni-IDA or Ni-NTA resin. The particular metal and chelation chemistry of a support determine its binding properties and suitability for specific applications of IMAC.

Affinity purification of His-tagged fusion proteins is the most common application for metal-chelate supports in protein biology research. Nickel or cobalt metals immobilized by NTA-chelation chemistry are the systems of choice for this application. In addition, different varieties of agarose resin provide supports that are ideal for His-tagged protein purification at very small scales (96-well filter plates) or large scales (series of chromatography cartridges in an FPLC system). When packed into suitable columns or cartridges, resins such as Ni-NTA Superflow Agarose provide for purification of 1 to 80 milligrams of His-tagged protein per milliliter of agarose beads. Compared to cobalt and other ligands used for IMAC, nickel provides greater capacity for His-tagged protein purification. Thermo Fisher Scientific offers HisPur Ni-NTA Superflow Agarose that exhibits a high dynamic binding capacity across a range of flow rates, making it an excellent choice for larger scale purifications.

High-yield, high-purity, medium-scale purification of 6xHisTagged protein. More than 4 grams of over-expressed 6xHis-GFP were purified in 3 hours using 200 mL columns containing HisPur Ni-NTA Superflow Agarose. One liter of lysate was loaded at a flow rate of 20 mL/min, then washed until baseline with wash buffer containing 30 mM imidazole. Bound protein was eluted with buffer containing 300 mM imidazole. Fractions containing purified 6xHis-GFP were pooled and quantitated using Pierce 660 nm Protein Assay. Load, flow-through, wash, and elute fractions were separated by SDS-PAGE, stained with Imperial Protein Stain.

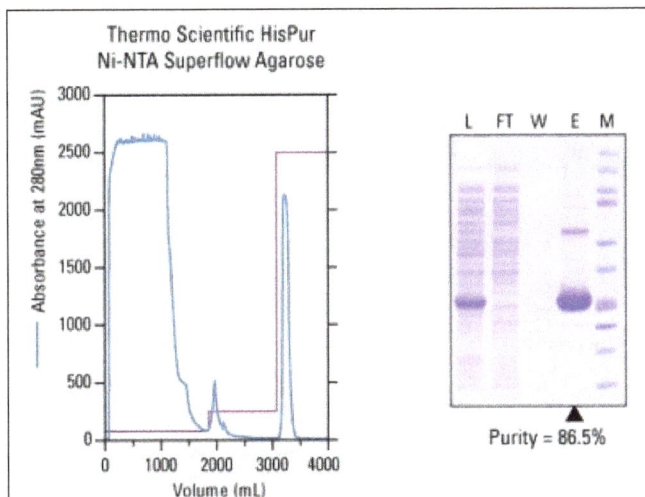

Thermo Scientific HisPur
Ni-NTA Superflow Agarose

Purity = 86.5%

Poly-His tags bind best to IMAC resins in near-neutral buffer conditions (physiologic pH and ionic strength). A typical binding/wash buffer consists of Tris-buffer saline (TBS) pH 7.2, containing 10-25 mM imidazole. The low-concentration of imidazole helps to prevent nonspecific binding of endogenous proteins that have histidine clusters. (In fact, antibodies have such histidine-rich clusters and can be purified using a variation of IMAC chemistry.)

High concentrations of salt and certain denaturants (e.g., chaotropes such as 8 M urea) are compatible, so purification from samples in various starting buffers is possible. For this reason, it is best to use the His-tag for design and expression of recombinant proteins that may need to be purified in denatured form from inclusion bodies. (Contrast this with the GST-tag, which is an enzyme that must remain functional to enable purification.) However, reducing agents, oxidizing agents and chelators (e.g., EDTA) are generally not compatible with IMAC affinity chemistry.

Elution and recovery of captured His-tagged protein from an IMAC column is accomplished by using a high concentration of imidazole (at least 200 mM), low pH (e.g., 0.1 M glycine-HCl, pH 2.5) or an excess of strong chelators (e.g., EDTA). Imidazole is the most common elution agent.

Be aware that immunoglobulins are known to have multiple histidines in their Fc region and can bind to IMAC supports. High background and false positives can result if binding conditions are not sufficiently stringent (i.e., with imidazole) and the immunoglobulins are abundant relative to the His-tagged proteins of interest. Albumins, such as bovine serum albumin (BSA), also have multiple histidines and can bind to IMAC supports in the absence of His-tagged proteins in the sample or imidazole in the binding/wash buffer.

Thermo Scientific HisPur Cobalt Resin is a tetradentate chelating agarose resin charged with divalent cobalt (Co^{2+}). The resin provides a high degree of purity and may recover more than 10 mg of pure His-tagged protein per milliliter of resin without metal contamination or the need to optimize imidazole washing conditions.

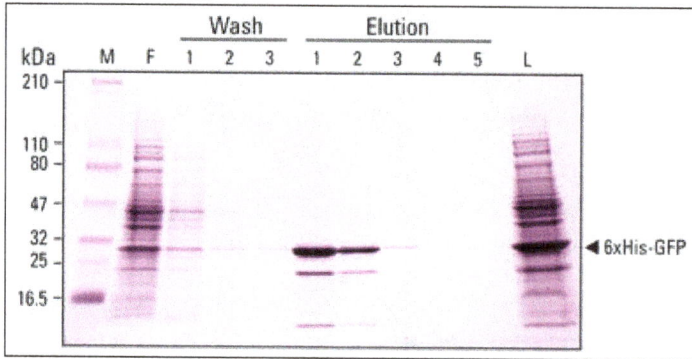

Affinity purification of His-tagged proteins. Cell lysate containing over-expressed recombinant 6xHis-tagged Green Fluorescent Protein (GFP) was prepared in B-PER Bacterial Protein Extraction Reagent and protease inhibitors. Protein concentrations were determined by Coomassie Plus Protein Assay. Bacterial lysate (1.0 mg total protein) was applied to a 0.2 mL bed volume of HisPur Cobalt Resin in a spin column. The resin was washed three times with 0.4 mL of wash buffer containing 10 mM imidazole. His-tagged proteins were eluted three times with 0.2 mL of elution buffer containing 150 mM imidazole. Gel lanes were normalized to equivalent volume. Gel was stained with Imperial Protein Stain M = Molecular Weight Marker; L = lysate load; FT = flow-through.

Nickel, Cobalt and Copper

Nickel is the most widely available metal ion for purifying His-tagged proteins. Nickel generally provides good binding efficiency to His-tagged proteins but also tends to bind nonspecifically to endogenous proteins that contain histidine clusters. As stated above, a small amount of imidazole in the binding/wash buffer helps to control off-target binding.

Cobalt exhibits a more specific interaction with histidine tags, resulting in less nonspecific interaction. For this reason, cobalt is the preferred divalent cation for purifying His-tagged proteins when high purity is a primary concern. Thermo Fisher Scientific offers both nickel and cobalt IMAC resin formats.

Ni-NTA and Cobalt resins maximize yield and purity, respectively. Gel panels: Bacterial lysate containing over-expressed 6xHis-AIF2 (6 mg total protein) or 6xHis-GFP (4 mg total protein) was applied to HisPur Ni-NTA Resin (0.2 mL) and purified by the batch-bind method. The same amount of total protein was applied to Ni-IDA and HisPur Cobalt resins and purified according to the manufacturer's instructions. All Gels lanes were normalized to equivalent volume. M = molecular-weight marker, L = lysate load and FT = flow-through.

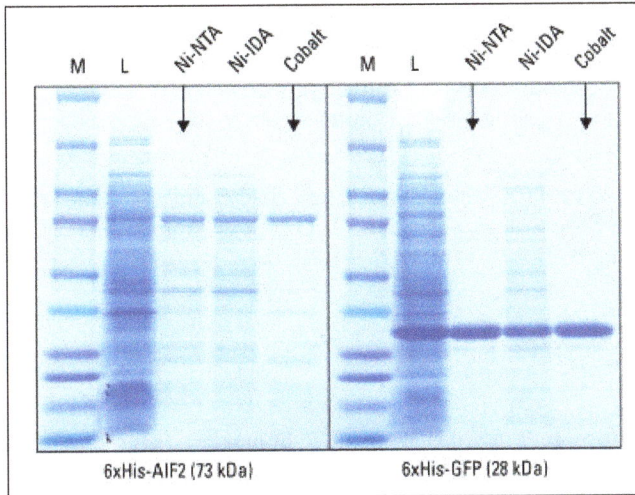

6xHis-AIF2 (73 kDa)   6xHis-GFP (28 kDa)

Copper ions bind his-tags more strongly than cobalt or nickel. This provides the highest possible binding capacity but also the poorest specificity. For this reason, copper IMAC is commonly used only for binding applications in which purification is not the objective (e.g., plate-coating of an already-purified His-tagged protein for use in an assay).

Other His-tagged Fusion Protein Techniques

Besides affinity purification, other applications for His-tagged fusion proteins are made possible with the aid of IMAC-type chemistries or His-tag-specific antibodies:

- Microplate coating: Nickel - or copper-coated microplates enable fusion proteins to be coated from crude or semi-purified samples for plate and reporter assays of various kinds.

- ELISA or Western blot detection: Nickel-chelated horseradish peroxidase (Thermo Scientific HisProbe HRP) enables HRP-based detection of His-tagged proteins without antibodies. Alternatively, anti-6xHis antibodies are also available.

- Protein interaction pull-down: Nickel agarose resin can be used to purify, identify and measure interactors of His-tagged proteins.

- Gel staining: A metal-based fluorescent stain enables detection of His-tagged proteins in SDS-PAGE.

References

- Protein-quantification, biopharmaceuticals, pharmaceutical: intertek.com, Retrieved 19 June, 2019

- Dyson HJ, Wright PE (March 2005). "Intrinsically unstructured proteins and their functions". Nature Reviews. Molecular Cell Biology. 6 (3): 197–208. Doi:10.1038/nrm1589. PMID 15738986

- Protein-structure, technical-briefs, news: particlesciences.com, Retrieved 18 May, 2019

- "Definition: expression system". Online Medical Dictionary. Centre for Cancer Education, University of Newcastle upon Tyne: Cancerweb. 1997-11-13. Retrieved 2008-06-10

- His-tagged-proteins-production-purification, pierce-protein-methods, protein-biology-resource-library, protein-biology-learning-center, protein-biology, life-science, home: thermofisher.com, Retrieved 20 July, 2019

- Bu Z, Callaway DJ (2011). "Proteins move! Protein dynamics and long-range allostery in cell signaling". Protein Structure and Diseases. Advances in Protein Chemistry and Structural Biology. 83. Pp. 163–221. Doi:10.1016/B978-0-12-381262-9.00005-7. ISBN 9780123812629. PMID 21570668

- Gst-tagged-proteins-production-purification, pierce-protein-methods, protein-biology-resource-library, protein-biology-learning-center, life-science, home: thermofisher.com, Retrieved 21 August, 2019

- PDB: 2RR8; Umemoto R, Nishida N, Ogino S, Shimada I (September 2010). "NMR structure of the calponin homology domain of human IQGAP1 and its implications for the actin recognition mode". J. Biomol. NMR. 48 (1): 59–64. Doi:10.1007/s10858-010-9434-8. PMID 20644981

- Domains-ferm, ferm-protein-domain, resources-protein-domains-interactions, contents: cellsignal.com, Retrieved 22 January, 2019

Branches of Proteomics

There are various branches of proteomics such as structural proteomics, interaction proteomics, phosphoproteomics, imunoproteomicsm, secretomics, activity-based proteomics, neuroproteomics and quantitative proteomics. This chapter closely examines these branches of proteomics to provide an extensive understanding of the subject.

STRUCTURAL PROTEOMICS

Structural proteomics is the process of the high-throughput characterization of the three-dimensional structures of biological macromolecules.

One of the most spectacular recent achievements in life sciences has been the sequencing of the entire human genome, accomplished by the Human Genome Project. The resolution of the entire sequence of the human genome has resolved many unanswered questions relating to human life. The human body comprises a vast number of cells and each cell contains many thousands of different proteins necessary to maintain cellular function. Knowledge of the sequence of the human genome means that disease-associated abnormalities can now be detected at the genetic level. Furthermore, sequence comparisons can provide an insight into the evolutionary relationship between organisms.

As of August 2011, the UniProtKB/Swiss-Prot database has contained in excess of half a million non-redundant sequence entries. Hence, it is clear that large-scale genomic projects have provided the sequence infrastructure for the in-depth analysis of proteins. A new fundamental concept of the proteome (Protein complement to a Genome) has emerged that aims to unravel the biochemical and physiological mechanisms of complex multivariate diseases at the functional and molecular level. As a consequence, the new science of proteomics has been established to complement physical genomic research. Proteomics can be defined as the quantitative comparison of proteomes under different conditions, which aims to further characterize biological processes and functional protein networks However, the knowledge gleaned from the various genomes sequenced to date is not sufficient to understand the function of proteins within the cell. To characterize functional protein networks and their dynamic alteration during physiological and pathological processes, proteins have to be identified, sequenced, categorized and classified with respect to their function and interaction partners. To understand their functions at a molecular level, it is often necessary to determine their three-dimensional (3D) structures at atomic resolution.

During the past decade, the emerging field of structural proteomics (SP) has developed, representing an international effort aimed at the large-scale determination of the 3D structures of proteins encoded by the genomes of key organisms. Initiatives in SP research have led to the development of novel strategies and automated protein structure determination pipelines around the world.

When protein structure analysis was first established in the late 1960s and the X-ray structures of myoglobin and hemoglobin were determined, the development of such a high-throughput (HT) infrastructure for protein structure analysis would have seemed like an impossible dream. The remarkable success and technological advancements since then have had a tremendous impact on throughput in protein structure determination and all stages of the pipeline have become more or less automated. Currently, SP initiatives are generating protein structures at an unprecedented rate and have resulted in an exponential growth in the number of protein structures deposited in the Protein Data Bank. However, the number of solved protein structures in the PDB represents only a small proportion of the theoretical number of proteins encoded by genomic sequences.

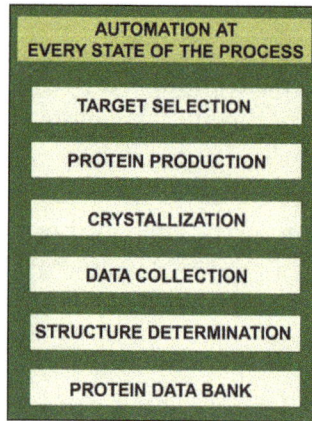

Process involved in SP using X-ray crystallography.

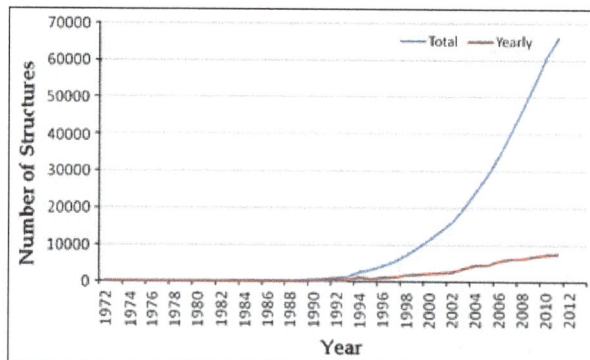

Exponential growth in the number of X-ray protein.
structures deposited in the Protein Data Bank

To bridge this gap and to meet the demand of rapidly obtaining protein structure information, advancements have been made in SP methodologies in the form of HT technologies. However, these technologies have encountered some of the traditional bottlenecks in structure determination for difficult proteins and complexes of proteins at HT. To overcome these bottlenecks, efforts have been focused on improving the structure determination pipeline by streamlining and optimizing protein production, protein crystallization, data collection and structure solution. In addition, SP centers have adopted bioinformatics analysis of potential targets to generate models based on solved structures and to establish collaborative research to exploit the function of proteins. Recently, in the USA, the National Institutes of Health established a Protein Structure Initiative(PSI): a biology network to determine protein structures including membrane proteins of high biological interest. The objective of the PSI is to develop suitable technologies for membrane protein structure solution, using bioinformatics and modeling to leverage solved structures, and to carry out collaborative research to provide a link between a structure and its biomedical and biotechnological impact. On the other hand, in Europe, the emphasis for the Structural Proteomics IN Europe(SPINE) initiative has been to apply these HT technologies to systems of biological interest, the ultimate aim being to solve significant biological problems more effectively. Furthermore, the European INSTRUCTproject offers scientists access to world-class structural biology and SP infrastructures and expertise. INSTRUCTmakes integration possible more rapidly, creating a coherent forum for structural biology. This forum will stimulate closer collaboration between scientific communities and initiatives in biological sciences.

Automation and Strategies for Protein Structure Analysis

Protein Production and Crystallization

Generating pure, soluble and homogeneous protein for structure determination is a major rate- imiting step in the overall process. Traditional sequential generation of single expression constructs for a single protein target has been superseded by parallel, HT cloning techniques. Genetic engineering and the use of specific crystallization chaperones are two approaches that have proven invaluable for the determination of many highly important protein structures.

Screening of Candidate Proteins

Structural biology projects are typically initiated to characterize the biological activity of a specific protein or protein complex. In some cases, crystallization of that exact protein leads to a structure that can be correlated directly to functional data. However, in many cases, researchers will eventually come to the conclusion that the protein of interest is not suitable for structural analysis. It makes sense, therefore, to include parallel, HT approaches early on for identifying optimal boundaries and experimental conditions for protein production and crystallization. The selection of candidate proteins is very much project dependent, but will usually include orthologs or homologs of the original protein of interest and genetic constructs corresponding to subregions or individual domains. Methods for HT characterization of larger numbers of expression clones have originally been described for bacterial expression systems. Small-scale expression testing is more difficult to achieve in eukaryotic systems such as yeast and baculovirus. Transient transfectionof mammalian cell lines such as HEK293 is a highly efficient system for secreting mammalian glycoproteins and has also been successfully applied to produce membrane proteins such as rhodopsin. This method can be performed in a HT manner for characterization of protein candidates for crystallization.

Glycoproteins

The choice of expression system has a great influence on the quality and quantity of the produced recombinant protein. Cell-freeprotein production has proven its value for producing soluble and membrane proteins for NMR and crystallographic studies. Mammalian proteins stabilized by disulfide bonds and modified by glycosylation are especially demanding targets. Mammalian cells are the ideal host for these proteins, since they yield protein with all the post-translational modifications required for biological activity, including authentic glycosylation and correct disulfide pairing. However, cell culture is time and labor intensive. Many extracellular mammalian proteins can be recovered in active form through refolding of bacterial inclusion body proteins. Obviously, these proteins do not require post-translational modifications other than disulfide bridges for correct folding. Refoldingof proteins from inclusion bodies is common in industrial production but requires extensive process optimization. Structural biology projects applying inclusion body refolding benefit from automated screening of folding conditions with generic, biophysical assays.

Animal cell linesare highly effective for the secretion of proteins with native glycosylation and disulfide bonds. Glycoproteins produced with the mammalian CHO or HEK293 cell lines carry heterogeneous, complex-type oligosaccharide chains attached to Ser/Thr (O-linked) or Asn (N-linked) side chains. Crystallization of glycoproteins is difficult because of the heterogeneity and flexible conformation of the bulky oligosaccharides, which can also mask possible sites of crystal contacts on the protein surface. Some glycosylation sites can be removed by mutagenesis. Regions with O-linked glycosylation are generally proline-rich and unfolded, and can be excluded from genetic constructs. However, many proteins require glycosylation for folding and transport through the secretory pathway. Enzymatic removal of N-linked glycans from the purified protein with endoglycosidase H or F leaves a single monosaccharide attached, which may increase the solubility of the deglycosylated protein. Enzymatic deglycosylation is efficient for oligosaccharides of the high-mannose type as obtained from the baculovirus system. Processing of N-linked glycans by mammalian cell lines results in complex-type oligosaccharides that are difficult to cleave enzymatically. Complex-type glycosylation can be prevented by chemical glycosylation inhibitors or by mutating the host cells. The gene for the enzyme N-acetylglucosaminyl-transferase I (GnTI), which modifies high-mannose type oligosaccharides, has been mutated in the cell lines CHO

Lec1, Lec3.2.8.1 and HEK293S-GnTI(−). These cell lines and normal HEK293 cells treated with the glycosylation inhibitors kifunensine or swainsonine have enabled the production of many glycoproteins and their crystallization upon enzymatic deglycosylation. Optimized protocols and cell lines allow performing transient transfection of HEK293 at up to liter scale with inexpensive reagents. However, not all proteins can be produced in sufficient amounts by transient transfections. Stable cell lines allow the production of proteins more reproducibly and in much larger volumes in bioreactors. However, establishing lines with good performance requires considerable effort. Novel approaches of stable cell linedevelopment, based on preparative cell sorting and recombinase-mediated cassette exchange, combine faster development times with improved performance and have been used successfully for X-ray crystallography studies.

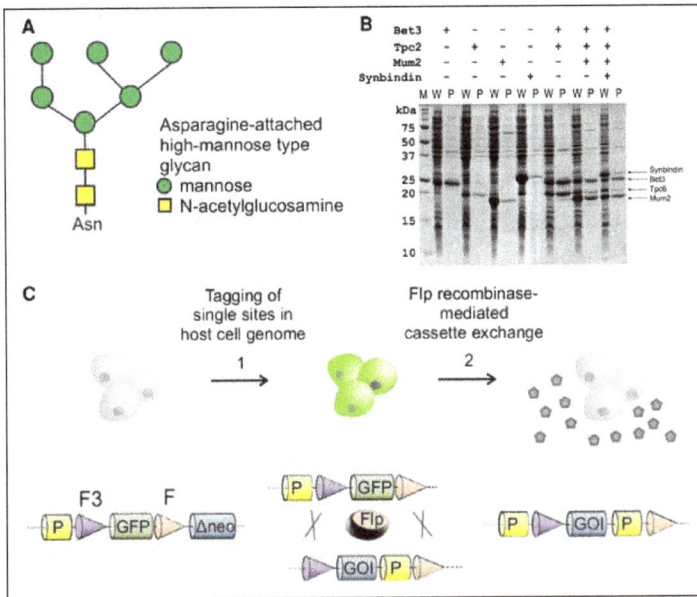

Protein production. aGlycosylation: structure of a high mannose-type glycan. bCo-expression of a complex of four proteins with the pQLink system.M: marker,W: whole cellular protein,P: purified protein. cCell line development by recombinase-mediated cassette exchange(RMCE): cells are transfected with a vector containing aGFPgene flanked by recombination sitesF3andFand GFP-positive cells are isolated. Cassette exchange is initiated by co-transfecting a tagged cell line with an Flp recombinase expression vector and a targeting vector bearing the gene of interest (GOI). Flp recombinase exchanges the tagging gene cassette and a production cell line is obtained.

Crystallization Chaperones

Some protein families require a combination of specialized strategies for successful crystallization. Crystallization chaperones are proteins that specifically bind to the target protein and support "carrier-driven" crystal growth. They limit the conformational flexibility of the target protein and provide a large, hydrophilic interaction surface for initiating crystal lattice contacts. Fab fragmentsof monoclonal antibodies have been used traditionally as crystallization chaperones. In addition, recombinant antibodies from camels (VHHs, also called "nanobodies") and synthetic scaffold proteins have demonstrated their usefulness in many examples. Disulfide-free synthetic scaffold proteins such as designed ankyrin repeat proteins(DARPINs) or the fibronectin type III domain (FN3) can be screened for specific binders in vitro and can be produced easily in E. coli. Fab fragments enabled the first crystal structure to be determined for a non-rhodopsin GPCR and a full-length potassium channel. Furthermore, the first high-resolution crystal structure of the β2 adrenergic receptor–Gs protein complex has recently been reported in which nanobodies (camelid antibody fragments) were used to significantly improve crystal quality. Nanobodies are relatively simple proteins, about a tenth the size of antibodies and just a few nanometers in length.

Protein Engineering

Protein engineering can overcome problems with producing sufficient amounts of protein, keeping protein soluble at the concentrations required for crystallization and obtaining proteins with surfaces that allow crystal formation. In general, protein engineering follows one of three strategies: designing shortened proteins lacking terminal residues outside the globular fold, mutating residues on the target protein's surface or designing fusion proteins.

If a full-length protein cannot be produced or crystallized, then a common strategy is to design shorter variants which represent isolated domains, eliminating flexible regions at the termini or large internal loops. Databases such as PFAMprovide information on the presence of conserved domainsin protein sequences. Alternatively, genetic constructs can be designed that avoid regions predicted to be disordered and unfolded by software tools such as DISOPRED2 , RONN, or the meta-server metaPrDOS. The strategy of designing genetic constructs on the basis of computational analysis may fail because of imprecise or missing information in the respective databases. Robotic screening of random truncation libraries represents an alternative technique in such cases. In this technique, the cDNA is fragmented and a library of expression clones is created by cloning the fragments. By chance, a few clones of the library will contain a fragment that encodes for just one complete domain. Such clones will express soluble protein. Different ways of screening libraries for clones that express soluble protein have been described, including a filtration technique, the biotinylation assay of the ESPRITtechnology and screening based on GFP fusion protein fluorescence. The ESPRIT(Expression of Soluble Proteins by Random Incremental Truncation) library technology has been adapted recently to allow screening for soluble protein complexes.

Single point mutations can have a dramatic effect on a protein's solubility and crystal formation. The most successful point mutation strategy, surface entropy reduction(S-ER), replaces small clusters of two to three surface residues with high conformational entropy such as Lys, Glu or Gln with Ala. SER produces mutants that are often more susceptible to crystallization than the wild-type protein. More than 100 structures of proteins optimized by SER have been solved. A Web server facilitates protein engineering for SER. In general, SER does not improve protein solubility. Proteins that cannot be concentrated to sufficient levels for crystal growth benefit from strategies opposite to SER. The solubility of such proteins can often be increased by reducing the hydrophobicity of surface residues. This approach is more difficult than SER, because exchanging hydrophobic surface residues requires some knowledge of the protein's structure.

Directed point mutations are generally not used to improve the protein production levels since no rational strategies are available. However, screening of large libraries of random mutants of the target proteins, enabled by laboratory automation, has been successful. Fusion proteins with partners such as glutathione S-transferase (GST), thioredoxin or maltose binding protein (MBP) are a more common approach to improve the target protein's production and solubility. However, the flexible linker between the fusion partners generally inhibits crystallization. Furthermore, when the fusion partner is removed using a site-specific protease, the improvement in solubility conferred by the fusion partner may be lost. Careful design of MBP fusion proteins enables carrier-driven crystallizationof intact fusion proteins. These fusions have to be designed in such a way that the MBP's C-terminal α-helix is fused directly to the globular core of the target protein, thereby avoiding flexibility between the fusion partners. Then, the MBP part can improve protein yield and solubility and promote crystal growth.

One method of surface modification that does not involve additional cloning is reductive lysine methylation, where lysine side chains are chemically modified. The technique can improve the X-ray diffraction of existing crystals, or permit the crystallization of proteins that had previously failed to yield crystals.

Protein Complexes

Protein complexes are attractive targets for X-ray crystallography, because their structures reveal important information relating to the molecular details of specific protein recognition. However, crystallization of a complex requires careful preparation that includes critical assessment of the available data, careful optimization of sample preparation and functional and biophysical characterization of the complex using a variety of methods. Very stable complexes that do not dissociate are preferred targets for crystallization. However, the subunits of such stable heterocomplexes may not be able to fold into a soluble conformation alone, necessitating the co-expressionof the complex components. Transient complexes, on the other hand, which exist in equilibrium with the dissociated subunits, are more difficult to crystallize because of sample heterogeneity. Subunits of transient complexes may form crystals that exclude the other subunit, which is often difficult to detect. Recombinant production of the subunits of a protein complex in the same host cell by co-expression has been described with a large variety of systems. Novel cloning strategies enable co-expression of many subunits in host cells including E. coli, baculovirus and mammalian cells, and have been adapted to automated cloning. The pQLinksystem allows co-expression of an unlimited number of protein subunits in E. coliwith different affinity tags. pQLinkvectors have been widely used by different laboratories, mainly for eukaryotic vesicle tethering complexes. Studies comparing a large variety of expression systems have demonstrated that subtle changes in the expression strategy have a profound effect on the success of co-expression experiments, even if the main parameters, protein sequence and host cell are identical.

Successful recombinant expression of protein complexes requires that the subunits are synthesised in similar amounts. Otherwise, the yield of the complete complex is determined by the subunit present in the lowest concentration. Also, a heterogeneous mixture of the complete complex with smaller oligomers not comprising all subunits is obtained. To circumvent this problem, the synthesis of polyproteinshas been introduced for generating protein complexes. This strategy is reminiscent of the genomes of many viruses that contain large open reading frames encoding polyproteinsthat are cleaved by viral proteases into single proteins upon translation. A baculovirus vector containing a large open reading frame comprising single protein sequences separated by a site-specific protease site was created. The coding sequence of the TEV protease was included in the vector. Upon overexpression, intracellular TEV protease cleaved the polyprotein into single subunits of a protein complex. This strategy was successfully demonstrated for sub-complexes of human general transcription factor TFIID and other complexes.

Protein Crystallization Methods and Automation

Production of protein crystals suitable for structural studies poses one of the major bottlenecks in the entire process. Finding crystallization conditions that yield single, well-ordered crystals with low mosaicity that diffract to sufficient resolution can be very challenging. The quality of a crystal is often linked to the number of crystals formed (a few large crystals versus many microcrystals), size (larger is better) and appearance (optically clear, sharply faceted crystals are best). However, any true measure of quality must verify that the diffraction properties correlate with the morphological quality of the crystal.

Crystallization can occur spontaneously, or alternatively it can take several days, weeks or months for crystals to appear. Longer crystallization times are usually indicative of proteolytic cleavage at the protein termini promoting crystal formation. It is not easy to provide an estimate for maximum protein crystal growing time. There have been some reports showing that, in some cases, protein crystals may take as long as 6 months or a year to appear. However, an average growing time for a protein crystal is typically less than a month. Normally, protein crystallization occurs when the concentration of protein in solution is greater than its limit of solubility, so that the protein solution becomes supersaturated. To crystallize a protein, it undergoes slow precipitation from an aqueous solution. As a result, individual protein molecules align themselves in a repeating series of "unit cells" by adopting a uniform orientation. One unavoidable aspect of crystallizing a newly expressed protein is the need to carry out a large number of experiments to find suitable conditions in which the protein crystallizes. It can be extremely tedious and time consuming to set up a broad array of different crystallization experiments manually. With the advent of HT liquid handling and crystallization systems, it is relatively easy to prepare a thousand or more crystallization experiments in which crystallization parameters, such as the ionic strength, pH, protein and precipitant concentration and temperature, are varied systematically. However, the success rate does not depend upon the number of crystallization conditions tested.

Methods used for crystallization include vapor diffusion, batch crystallization, dialysis, seeding, free-interface diffusion and temperature-induced crystallization. The most popular method for setting up crystallization experiments is vapor diffusion, which includes hanging drop (for smaller volumes), sitting drop (for larger volumes), the sandwich drop, reverse vapor diffusion and pH gradient vapor diffusion methods. A drop containing a mixture of precipitant and protein solution is sealed in a chamber with pure precipitant. Water vapor subsequently diffuses from the drop until the osmolarity of the drop and the precipitant is equal. The dehydration of the drop causes a slow concentration change of both protein and precipitant until equilibrium is achieved, ideally in the crystal nucleation zone of the phase diagram. Batch crystallization relies on bringing the protein directly into the nucleation zone by mixing protein with the appropriate amount of precipitant. The batch method is usually carried out under oil to prevent the diffusion of water out of the drop. Many of these methods can be performed using HT automated instrumentation and miniaturization of crystallization experiments and have had huge impacts on protein crystallization in terms of saving time and conserving precious sample. For example, crystallization robots such as the Phoenix RE(Rigaku Corporation) and the Mosquito(TTP Labtech), which can accurately and reproducibly dispense very small volumes (nl in size) into 96-well plates for automated screening and optimization of crystallization conditions, are now commonplace in many laboratories.

In addition, TTP LabTech's MosquitoLCP(Lipid Cubic Phase) has been designed to aid in the crystallization of membrane proteins by accurately dispensing nanoliter quantities of highly viscous lipids or detergents that are required to retain the structural integrity of the sample. A recent development in protein crystallization has been the use of high-density, chip-based microfluidic systems for crystallizing proteins using the free-interface diffusion method at nanoliter scale, including Emerald Biosystems MPCS(Microcapillary Protein Crystallization System), Fluidigm Corporations TOPAZsystem and the Microlytic Crystal Former. These platforms have the advantage of using minimal protein sample to screen a broad range of crystallization conditions. The Rigaku CrystalMation system was set up to fully automate the crystallization process while dealing with sample volumes of 100 nl per experiment.

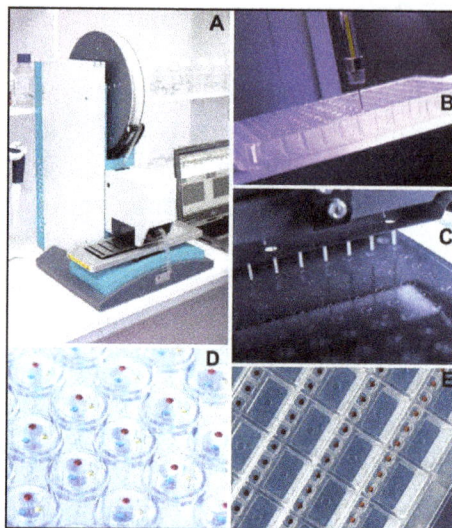

Protein crystallization and automation. aTTP LabTech's mosquito Crystal automates protein crystallography vapor diffusion set-ups, additive screening and microseeding; bTTP LabTech's mosquito LCP: a dedicated instrument for crystallising membrane proteins using lipidic cubic phase screening. The panel highlights the positive displacement syringe, which dispenses the highly viscous lipid mesophases used in the LCP technique into 96-well crystallization plates. (cand d) Crystallization plate set up for hanging drop vapor diffusion experiments; enanoliter sitting drop experiments set up in a 96-well plate.

A popular strategy for the optimization of crystallization conditions in vapor diffusion is crystal seeding. Seeding decouples nucleation from crystal growth and involves transferring previously obtained seed crystals into undersaturated drops. Homogeneous seeding techniques include microseeding, streak seeding and macroseeding. Seed stock for microseeding can be conveniently generated using Hampton Research's Seed Bead kit. More recently, a simple, automated microseeding technique based on microseed matrix screening has been developed. This method consists of the addition of seeds into the coarse screening procedure using a standard crystallization robot and has been shown to not only produce extra hits, but also generate better diffracting crystals. Successful cases for a simple semi-automated microseeding procedure for nanoliter crystallization experiments have also been recently described. Furthermore, crystallization plate storage and inspection are now fully automated. For example, the Minstrel drop imager family (Rigaku) and Rock Imager (Formulatrix) combine imagers with gallery plate hotels/incubators to store crystallization plates at a constant temperature, periodically inspect them and manage the data.

Despite the progress that has been made in increasing throughput, the act of identifying crystals in the crystallization experiments remains a task requiring human intervention. A number of attempts are being made to automate crystal detection from the imaged drop and varying degrees of success have been reported. Automated crystal recognition has the potential to reduce the time-consuming human effort for screening crystallization drop images. Several approaches have been suggested to increase contrast for imaging and detection of protein crystals in such cases: crystal birefringence , addition of fluorescent dyes and monitoring the fluorescence of trace labeled protein molecules. The identification of crystallization hits has been simplified by UV detection combined with conventional imaging. For example, the latest generation of imaging systems combine visible and UV inspections providing a powerful tool for monitoring crystallization trials: when crystals are still too small to be mounted, the intrinsic protein fluorescence signal gives confidence that a crystallization hit is worth pursuing. Second-order nonlinear optical imaging of chiral crystals(SONICC) is an emerging technique for crystal imaging and characterization. SONICC imaging has been found to compare favorably with conventional optical imaging approaches for protein crystal detection, particularly in non-homogeneous environments that generally interfere with reliable crystal detection by conventional means.

A recent development is the X-CHIP(X-ray Crystallization High-throughput Integrated Platform): a novel microchip that provides a stable microbatch crystallization environment and combines multiple steps of the crystallographic pipeline from crystallization to diffraction data collection onto a single device. This system facilitates HT crystallization screening, visual crystal inspection, X-ray screening and data collection. The chip eliminates the need for manual crystal handling and cryoprotection of crystal samples while allowing data collection from multiple crystals in the same drop.

Data Collection and Processing

Data acquisition involves the recording of a series of X-ray diffraction images using a detector. The process of crystal mounting, centering, exposing with X-rays, recording diffraction data and dismounting the crystal represent the major steps in crystallographic data collection.

Radiation source, crystal handling and detector past and present tools: In the past, protein crystals typically ranging in size from tenths of a millimeter to several millimeters were mounted in glass capillary tubes. To collect data, the capillary tube was mounted on a goniometer and exposed to X-rays at room temperature. These X-rays were generated by low flux, sealed tube sources. Nowadays, data collection is handled by automated sample changers and micro-diffractometers in a cryogenic (100 K) environment utilizing brighter synchrotron radiation as the X-ray source. Cryo-freezing the sample inhibits free radicals diffusing through the crystal during data collection: these free radicals cause secondary radiation damage that leads to degradation in the quality of collected data. There are currently in excess of 125 dedicated protein crystallography beamlines around the World. The X-ray films that were used for data recording in the past have now been superseded with charge-coupled devices (CCD) and pixel array detectors, which allow diffraction data to be recorded directly and stored straight to disk. For example, a recent development has been the PILATUSdetector (pixel apparatus for the SLS), which has no readout noise, superior signal-to-noise ratio, a readout time of 5 ms and high dynamic range compared to CCD and imaging plate detectors. Delivery of high flux beam at third-generation synchrotron sources coupled with the advances in detector technology and control systems have significantly accelerated the speed of macromolecular diffraction data collection. An example of a state-of-the-art synchrotron X-ray data collection setup is shown in figure. Nowadays, crystals larger than 50 μm in size can be evaluated at conventional synchrotron beamlines. However, with some targets such as membrane proteins and multi-protein complexes, it is notoriously difficult to obtain crystals of sufficient size and order to generate high-quality diffraction data. Hence, next generation microfocus beamlines with reduced beam sizes have been established at synchrotron sites around the world, allowing measurements to be made on crystals a few micrometers in size. It has been predicted that a complete data set with a signal-to-noise ratio of 2σ at 2 Å resolution could be collectable from a perfect lysozyme crystal measuring just 1.2 μm in diameter using a microfocus beam. A number of crystal structures have been solved using micrometer-sized crystals by merging data from several crystals, including a polyhedron-like protein structure (~5–12 μm) and a thermally stabilized recombinant rhodopsin (with crystal dimensions of $5 \times 5 \times 90$ μm³) . Recently, strategies have been developed to determine structures from showers of microcrystals using acoustic droplet ejection (ADE) to transfer 2.5 nl droplets from the surface of microcrystal slurries through the air and onto micromesh loops. Individual microcrystals are located by raster-scanning a several-micron X-ray beam across the cryocooled micromeshes. X-ray diffraction data sets are subsequently merged from several micrometer-sized crystals and this technique has been used to solve 1.8 Å resolution crystal structures.

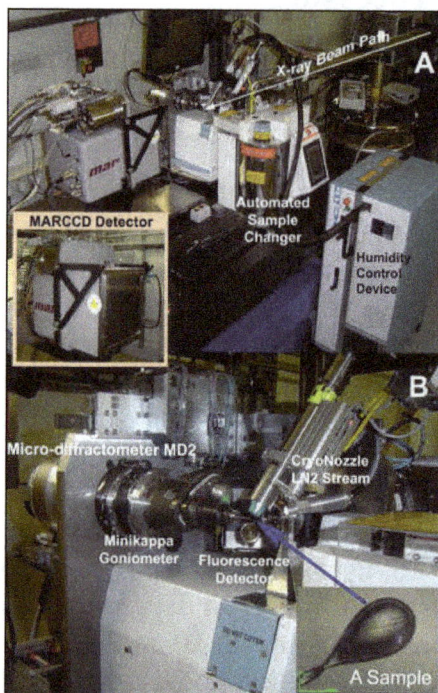

X-ray data collection facility. a End-station instrumentation at ESRF beam-line BM14 illustrating the sample changer used to exchange cry-frozen crystals on the potentiometer and the MARC (Mar-research GmbH) detector used to collect diffraction images.

The arrow high lights the path of the X-ray beam. b Close-up view showing the frozen crystal sample in the center of the image and the surrounding beam-line instrumentation. The red cross and blue circle-re present the center and diameter of the X-ray beam on the frozen crystal sample (bottom right).

As a result of these technological advancements, the time required to setup a diffraction experiment has become a significant proportion of the total time of an experiment. The diffraction experiment involves sample mounting, crystal centering and determination of data collection parameters. Significant progress has been made in automating crystal mounting, crystal centering and the energy scan to find metals or ions present in crystals that can be used for phasing. Automated sample mounting systems allow users to mount samples on the beamline without entering the experimental hutch. These systems minimize the need for manual intervention and facilitate the rapid and systematic screening of dozens of samples. For example, the automated sample changers equipped at the EMBL/ESRF beamlines are capable of handling 50 frozen samples, whereas the ACTOR(Automated Crystal Transfer, Orientation and Retrieval) robots installed on the beamlines at the DIAMOND synchrotron can mount up to 80 cryogenically frozen samples from their onboard storage dewars. This facilitates the rapid screening and ranking of crystals and enables users to collect data from their best diffracting crystal. These features make the automated approach far quicker than manual operations insuring that beamtime is used efficiently.

Before starting data collection, the crystal needs to be aligned so that it is coincident with the X-ray beam and the rotation axis. This is normally performed manually by the user at the beamline. However, for fully automated operation of the beamline, automated crystal centering is a prerequisite, especially when sample mounting robots are used. Semi-automated crystal centering based on a user clicking a mouse to indicate the position of the crystal through a specially designed software interface has been shown to be relatively robust and is employed at most synchrotron beamlines. Recent reports show that it is possible to center crystals automatically without user intervention using the recognition software C3D, XREC or alternatively by using the diffraction method. Crystal centering based on the diffracton method is especially attractive for micrometer-sized crystals. Optical centering of small crystals is challenging since visible light wavelengths (0.4–0.7 µm) are comparable to the crystal size and many crystals have irregular diffraction quality, which cannot be addressed by this technique. In diffraction-based crystal centering, the crystal is scanned in two dimensions using a small step size and at each step a diffraction image is taken, which is analyzed for locating and counting diffraction spots. The scored results are presented in a table which allows users to select optimally diffracting areas within the macroscopic sample.

Data Collection and Processing Software Packages

Typically, data extending to 2.5 Å resolution or higher are desirable for novel proteins and protein–ligand complexes, so that the model can be fitted unambiguously into the electron density map. However, in more challenging cases, data at 3 Å resolution or lower may be sufficient to fit the overall fold of a protein or the constituents of a multi-protein complex. A typical X-ray diffraction image, the electron density map to atomic resolution and the distribution of resolutions for protein structures in the PDB are depicted in figure. However, in many cases, diffraction properties of crystals are not known in advance, especially when crystals are small (in the micrometer range) and cannot be prescreened using in-house instrumentation prior to a synchrotron trip. It often takes a significant amount of time at the synchrotron to screen these sub-micron crystals to identify a well-diffracting crystal suitable for data collection. Whilst collecting data at the synchrotron beamline, the user must make decisions about the parameters of the experiment—exposure time, rotation range, oscillation angle, detector distance, beam size and wavelength—based on their experience, visual inspection of the diffraction images and information output by data-processing packages. Most of the instrumentation in the experimental station is computationally controlled using software packages such as Blu-Ice, CB-ASS, MxCube and J Blue-Ice. However, very often an intuitive decision is made by the user on the exposure time to use. In cases where this has been overestimated, it can lead to significant radiation damage before the completion of data collection. In addition, an inappropriate data collection strategy can lead to the failure of an experiment. Computationally efficient modeling of the data statistics for any combination of data collection parameters provides a foundation for making a rational choice. The modeling of data statistics using a few test images allows one to quantitatively select which screened crystal gives the highest resolution using an appropriate rotation range and X-ray radiation dose prior to data collection.

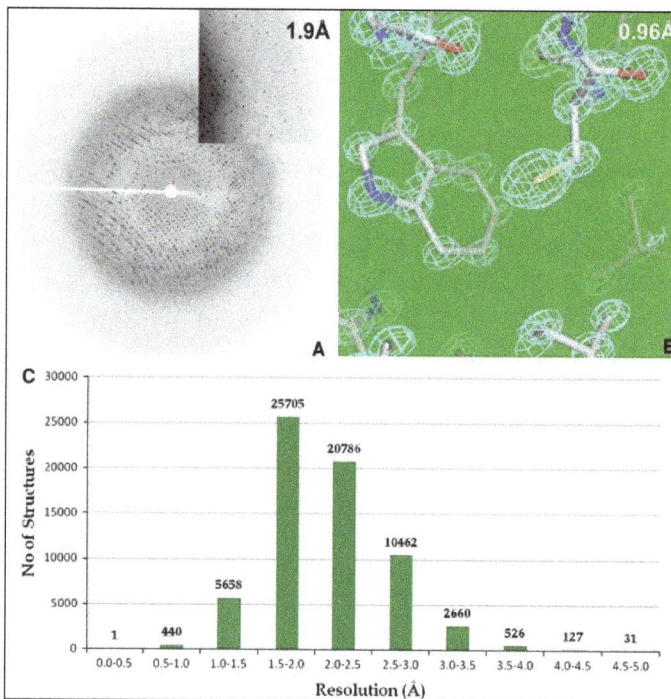

Accuracy and details. a Representative X-ray diffraction pattern collected on a Mar-research GmbH imaging plate system. The diffraction extends to a maximum of 1.9 Å resolution at the edge of the image. b Representative portion of an electron density map at 0.96 Å resolution. The sticks represent the individual atoms for the amino acids that constitute the protein (carbon, gray; nitrogen, blue; oxygen, red; sulfur, yellow) and the chicken wire represents the corresponding experimental electron density for these atoms. c Histogram depicting the distribution of resolutions for protein structures in the PDB as of August 2011.

The evaluation of the collected reflection intensities on the diffraction images involves the integration of the total intensity within all pixels of the individual spot profiles. The crystallographic program HKL 2000 is capable of carrying out data processing automatically. Other commonly used data-processing packages include XDS and MOSFLM. These programs all give excellent results with high-quality diffraction data, although their treatment of imperfect data differs owing to different approaches to indexing, spot integration and the treatment of errors. These programs can process data from a wide variety of modern area detectors from manufacturers including Mar-research, Rigaku/MSC, ADSC and MacScience. All these programs require crystallographers to make informative decisions and to input the correct experimental parameters to process the data successfully. There are ongoing activities at several synchrotron beamlines to develop expert systems that aim to automate the data collection strategy using the software BEST, RAD-DOSE, MOSFLM and XDS to reduce the time required to successfully collect high-quality X-ray data.

Post-crystallization Treatments to Improve the Quality of Diffraction. Among the biggest problems in macromolecular crystallography is the relatively weak diffraction power of protein crystals and their sensitivity to ionizing radiation damage. Cryogenic methods provide great advantages in macromolecular crystallography, especially when synchrotron radiation is used for diffraction data collection. Apart from reducing the problem with radiation damage and enabling the storage and safe transport of frozen crystals, there are a number of additional benefits. For example, cryo-freezing can be exploited to trap normally unstable intermediates in enzyme-catalyzed reactions to permit their characterization. In addition, cryo-freezing can dramatically improve diffraction properties by reducing thermal vibrations and conformational disorder within the crystal, provided that the crystal is amenable to freezing and a suitable cryoprotectant has been selected. Of primary practical importance is the decrease in secondary radiation damage in the crystal caused by the diffusion of free radicals, typically permitting a complete data set to be collected from a single crystal. Cryogenic data collection has allowed efficient phasing using multi-wavelength methods.

When a crystal of a biological macromolecule is cooled to cryogenic temperatures, the main difficulty is to avoid the crystallization of any water present in the system, whether internal or external. Therefore, a cooling procedure has to be chosen that leads to a glass-like amorphous phase of the solvent. In principle, there are four options: (1) cooling on a timescale too fast for ice formation to occur, (2) cooling at high pressure by which the formation of the common hexagonal form of ice is circumvented, (3) replacing the liquid surrounding the crystal with a water-immiscible hydrocarbon oil such as paratone-N, paraffinoil and LV CryoOil (MiTeGen), (4) modifying the physic o chemical properties of the solvent by the addition of cryoprotectants in a way that a vitrified state can be reached at moderate cooling rates.

To prevent the nucleation of ice crystals, the last method is currently the most widely used. The crystal is permeated with a diffusible solvent containing cryoprotectants such as glycerol, sucrose or other organic solvents. Determining the initial and optimal cryoprotectant concentration is often a process of trial and error. One must find suitable cryoprotectant concentrations that do not destroy the crystalline order while, at the same time, allowing the solvent to form an amorphous glass upon rapid cooling. Recently, trimethylamine N-oxide (TMAO) has been shown as a very versatile cryoprotectant for macromolecular crystals.

It has been shown that diffraction properties of flash-cooled macromolecular crystals can often be improved by warming and then cooling a second time—a procedure known as crystal annealing. Two different crystal-annealing protocols have been reported and many variants of these have been tried in the field. The first method involves removing a flash-cooled crystal from the cold gas stream and placing it in a cryoprotectant solution (either glycerol, MPD or Paratone-N oil) for several minutes before refreezing. There are several examples cited in the literature where this technique has been successfully applied. In the second method, the cold stream is blocked for a fixed amount of time before the crystal is allowed to re-cool. Both annealing protocols can improve crystal resolution and mosaicity, although substantial crystal-to-crystal and molecule-to-molecule variability has also been observed. Recently, the flash annealing technique has been automated using a cryo-shutter, a device that blocks the 100 K nitrogen stream that bathes the crystal for a specific amount of time. The main advantage of the shutter system is that it allows a controlled, instant re-cooling of the crystal and the user can perform the flash annealing experiment remotely without entering the experimental hutch.

Diffraction quality can also be improved by post-crystallization treatments, such as controlled dehydration, to attempt to improve the crystal diffraction properties. A user-friendly apparatus for crystal dehydration has been designed and implemented at the ESRF/EMBL beamlines. In addition, Proteros biostructures GmbH has developed a Free Mounting System(FMS) that precisely controls the humidity around a crystal, which can lead to dramatically improved diffraction data.

Remote Data Collection

Synchrotron data collection can be performed remotely from home institutions by accessing the instrumentation via advanced software tools that enable the network-based control of beamlines. Remote access to synchrotron sources is becoming more popular, since it saves both time and resource and results in more efficient use of the beamtime. "Mail-in" crystallography (diffraction data measured by synchrotron staff) is another popular option for X-ray diffraction data collection, whereby users ship their crystals to the synchrotron for data collection by the beamline scientists.

X-ray Structure Determination

Amplitudes or intensities can be measured directly from the X-ray diffraction experiment, but information relating to their relative phases cannot be measured. To be able to calculate an electron density map and subsequently determine the protein structure, an estimate of the phases has to be obtained indirectly using mathematical approaches and this represents the phase problemin protein crystallography.

Structure Determination Methods

Heavy-atom incorporation (isomorphous replacement, anomalous scattering and anomalous dispersion), molecular replacement and direct methods are commonly used techniques to solve protein structures. The general requirement for the exploitation of the anomalous signal for the determination of phase estimations via multipleor single-wavelength anomalous diffraction(MAD or SAD) techniques is that the protein crystal should contain anomalously scattering atoms, e.g., Hg, Pt or Se. With the advent of tunable X-ray sources and improved data collection techniques, it is now possible to measure the intensities of diffracted X-rays with very high precision. The small differences in intensities between Bijovoet pairs due to the presence of heavy atoms can be used to calculate initial estimates of the protein phase angle. One of the strategies widely used for the determination of novel protein structures is selenomethionine incorporation, where selenomethionine is replaced by methionine in the protein during expression. This method has revolutionized protein X-ray crystallography and it is estimated that over two-thirds of all novel crystal structures have been determined using either Se-SAD or Se-MAD. Novel structures can also be solved using the weak anomalous signals from atoms, such as sulfur and phosphorous present in certain macromolecules. SAD represents the most commonly used technique for novel proteins in SP centers. Multiple or single isomorphous replacement(MIR or SIR) methods also require the introduction of heavy atoms such as mercury, platinum, uranium or gold into the macromolecule under investigation.

These heavy atoms must be incorporated into protein crystals without disrupting the lattice interactions so that it remains isomorphic with respect to the native crystal. In the SIR method, intensity differences between the heavy-atom derivatized and native crystal are used to calculate experimental phases. Recently, the SIRphasing protocol has been re-applied in the radiation damage-induced phasing(RIP) technique, where the differences in intensities induced by radiation damage are used as a phasing tool. Limitations of these phasing protocols are mainly due to the deleterious effect that a high X-ray dose has on a protein crystal. X-ray radiation damage induces many changes to the protein structure and to the solvent, resulting in a consistent number of damaged sites and a decrease in the diffraction quality of the crystal. As an alternative to X-rays, ultraviolet(UV) radiation has been used to induce specific changes in the macromolecule, which only marginally affects the quality of the diffraction while inducing more selective changes to the protein structure.

This method is known as UV-RIP (ultraviolet radiation-damage-induced phasing). The most striking effect of UV radiation damage on protein crystals, as for X-ray radiation, is the breakage of disulfide bonds. Furthermore, this technique has been extended to a non-disulfide-containing protein, photoactive yellow protein, which contains a chromophore covalently attached through a thioester linkage to a cysteine residue and to selenomethionine (MSe) proteins. Therefore, this method offers considerable potential, and selenium-specific UV damage could serve as an additional or even an alternative way of experimental phasing in macromolecular crystallography. Another popular method adopted at SP centers is the use of iodide ion soaks and SA-Dexperiments for de novo phasing.

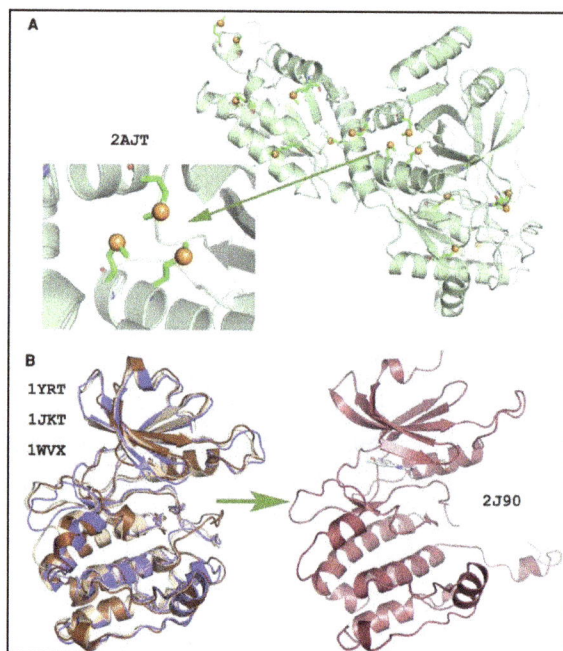

Examples for widely used structure determination methods. aStructure of E. coli-Arabinose Isomerase (PDB 2AJT) determined by single-wavelength anomalous diffraction(SAD). Selenomithionine residues are also shown. bStructure of DAPK3 (PDB 2J9o) determined with the molecular replacement (MR) method using the template prepared by homolog structures (PDB 1YRT, 1JKT, 1WVX).

Molecular replacement(MR) requires a search model for the protein under investigation, either determined from X-ray crystallography or from homology modeling, to calculate initial estimates of the phases of the new structure. The use of MR has become more commonplace with the expansion of the PDB and is currently used to solve up to 70% of deposited macromolecular structures where a homolog structure already exists.

Structural Proteomics for Biology

Structural knowledge of a protein clearly provides clues relating to its biological activity and physiological role. SP is one of the recent technologies that promotes drug discovery and biotechnological applications. Structural information can be used in many ways to ascertain the functional properties of cellular components. One of the crucial components for understanding the functions of novel proteins is the analysis of their experimental or modeled 3D structures. SP centers have provided an enormous impetus for methods development in structural biology and many laboratories are now actively implementing these technologies.

Systems Biology and Biotechnology

In systems biology, proteins are visualized as a network of interconnected dots. To understand the complexity of cellular function, one should know the detailed 3D behaviors of all the available dots which form the basis of life. Furthermore, the structures of these proteins could provide quantitative parameters to help elucidate functional networks through knowledge of protein function, evolution and interactions. The protein structures generated by SP can be used for the assignment of domain structure, functional annotation and the prediction of interaction partners in biochemical pathways. The structural information can be used to further characterize large-scale protein interaction networks by providing the key functional properties of cellular components.

Biotechnology embraces the bioproduction of fuels and chemicals from renewable sources. Sustainable energy is a major problem in the twenty-first century. If biofuels are to be part of the solution, this field must accept a degree of scrutiny unprecedented in the development of a new industry. That is because sustainability deals explicitly with the role of biofuels in insuring the well-being of our planet, our economy and our society, both today and in the future. The development of detailed kinetic models that include accurate regulatory network parameters will facilitate the identification of enzymatic bottlenecks in the metabolic pathways that could be harnessed to achieve biofuels overproduction. The latest advances in SP will continue to identify the biocatalysts, which power the development of enzyme reactors for producing substantial amounts of biofuels. Some biomolecules are robust enough to be used in biotechnological applications. For instance, enzymes can be used to break down starch to form sweeteners. The structure–function relationship of E. coliarabinose isomerase, ECAI, advanced its application in tagatose (a new sweetener) production.

Nanobiotechnology is a novel branch of futuristic science and engineering. A nanobiomachine is a machine formed by a biomolecule with a nanoscale diameter. The knowledge of a protein sequence provides the basis for understanding these nano-biomachines, which ultimately describe its functional significance. The structural and functional knowledge of a protein is essential to utilize proteins in nanotechnology applications and to develop bionanodevices. Glucose oxidase is a small, stable enzyme that oxidizes glucose into glucolactone, converting oxygen into hydrogen peroxide in the process. It is used as the heart of biosensorsthat measure the amount of glucose in the blood. Insights from protein structure can be crucial in engineering proteins for nanotechnology applications.

Models of Protein Structures and Drug Design

SP projects around the globe were established to determine the structures of proteins in an HT, automated fashion. However, despite the advances made by SP organizations in terms of automation, throughput and methodology development, the structures of certain classes of proteins, such as membrane proteins, are still notoriously difficult to determine. This warrants alternative techniques to generate models for these proteins to enhance our understanding of their physical and chemical properties. This has led to the development of a large number of bioinformatics tools capable of generating models for these novel proteins. Among them, MODELLER and ROSETTA represent two of the best protein structure prediction servers. MODELLE Rgenerates a model of an unknown protein using a template structure generated by the SP approach, whereas ROSETT Aprovides ab initio structure prediction of the unknown protein. Biomodeling provides the ability to understand the physicochemical properties of proteins of biomedical importance with undetermined 3D structures. It involves a range of computerized techniques based on quantum physics and experimental proteomics data to predict and correlate biological properties at the molecular level. Statistical and regression analysis techniques are the best methodologies and are capable of predicting geometries, energies, and electronic and spectroscopic properties. Homology-based modeling, as applied by MODELLER and ROSETTA, relies on sequence alignments between proteins of known structure and the target protein. The accuracy of the calculated model is dependent on the accuracy of the sequence alignment and the divergence between target and template. The most accurate alignments are obtained by iterative and profile or HMM(Hidden Markov Models)-based methods. In addition, structural data can be used to verify and improve alignments.

Biomodeling has become a valuable and essential tool in the drug design and discovery process. Drug design is a 3D puzzle where small drug molecules are fitted into the active site of a protein. The factors which affect protein–ligand interactions can be characterized by molecular docking and studying quantitative structure activity relationships (QSAR). Traditionally, drug discovery relies on a stepwise synthesis and screening of large numbers of compounds to optimize drug activity profiles. The design of new and more potent drugs against diseases such as cancer, AIDS and arthritis can be aided using bioinformatics tools such as computer-assisted drug design (CADD) or computer-assisted molecular design (CAMD). Structural bioinformatics tools not only have the potential to build predictive models of the proteins of biomedical interest, but also help to bring new drugs to market. Complementary in silicomethods, such as structure-based drug design (SBDD), incorporate the knowledge from high-resolution 3D protein structures generated by SP to probe structure–function relationships, identify and select therapeutically relevant targets (assessing druggability), study the molecular basis of protein–ligand interactions, characterize binding pockets, develop target-focused compound libraries, identify hits by HT docking (HTD) and optimize lead compounds, all of which can be used to rationalize and increase the speed and cost-effectiveness of the drug discovery process. An analysis of the results obtained by several docking and modeling programs has shown that, in most cases, they can work well. Most of the programs used in drug discovery have incorporated subroutines to identify false positives or negatives using scoring functions, which has led to a significant improvement in hit rates.

INTERACTION PROTEOMICS

Interaction Proteomics facilitates the study of protein-protein interactions (PPIs). Dysregulation and inappropriate activation of protein complexes are associated with several diseases, including cancer. These interactions play a vital role in cellular processes and fundamental biological insights can be gained by studying networks of interacting proteins rather than individual proteins.

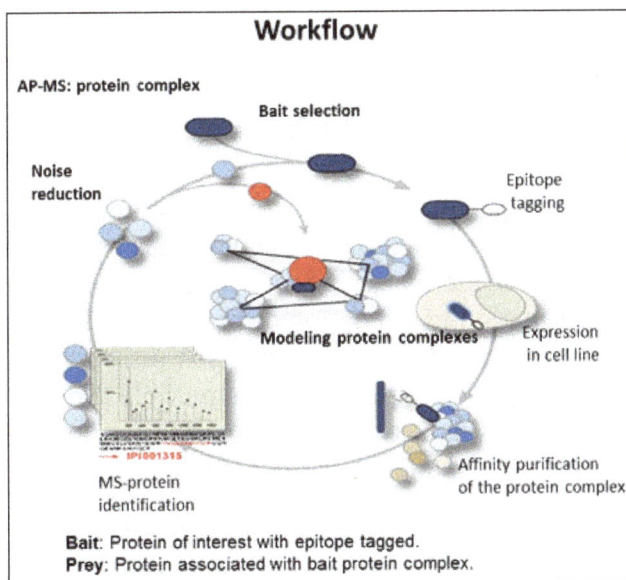

Workflow

AP-MS: protein complex

Bait selection

Noise reduction

Epitope tagging

Modeling protein complexes

Expression in cell line

MS-protein identification

Affinity purification of the protein complex

Bait: Protein of interest with epitope tagged.
Prey: Protein associated with bait protein complex.

We use the Affinity Purification-Mass Spectrometry (AP-MS) platform to study protein interactions and complexes. AP-MS combines the specificity of antibody based protein purification with the sensitivity of mass spectrometry to identify and quantify putative interacting proteins. Either native antibody or an epitope-tagging approach may be used to purify protein complexes. In the epitope-tagging approach, a recombinant epitope-tagged bait protein of interest is expressed in cultured cells, and then the bait protein and associated proteins are then retrieved using an antibody against the epitope, followed by identification of prey proteins using mass-spectrometry. The mass-spectrometry data is analyzed and scoring metrics applied to identify those proteins that are specific to the bait of interest. AP-MS enables protein networks and complexes to be studied at the proteomic scale and it can be anticipated that the knowledge gained from such projects will fuel drug target discovery and prove indispensable to the emerging field of systems biology.

Advantages

- Can provide functional information based on the function of proteins that are interacting with the bait.

- Provides information about protein molecular environment.

- Can help reconstruct pathways.

Disadvantages

- Up-front effort in constructing epitope-tagged baits.

- Care must be taken to exclude non specific or promiscuously-binding proteins.

PHOSPHOPROTEOMICS

Phosphoproteomics is a branch of proteomics that identifies, catalogs, and characterizes proteins containing a phosphate group as a posttranslational modification. Phosphorylation is a key reversible modification that regulates protein function, subcellular localization, complex formation, degradation of proteins and therefore cell signaling networks. With all of these modification results, it is estimated that between 30%–65% of all proteins may be phosphorylated, some multiple times. Based on statistical estimates from many datasets, 230,000, 156,000 and 40,000 phosphorylation sites should exist in human, mouse, and yeast, respectively.

Compared to expression analysis, phosphoproteomics provides two additional layers of information. First, it provides clues on what protein or pathway might be activated because a change in phosphorylation status almost always reflects a change in protein activity. Second, it indicates what proteins might be potential drug targets as exemplified by the kinase inhibitor Gleevec. While phosphoproteomics will greatly expand knowledge about the numbers and types of phosphoproteins, its greatest promise is the rapid analysis of entire phosphorylation based signalling networks.

A sample large-scale phosphoproteomic analysis includes cultured cells undergo SILAC encoding; cells are stimulated with factor of interest (e.g. growth factor, hormone); stimulation can occur for various lengths of time for temporal analysis, cells are lysed and enzymatically digested, peptides are separated using ion exchange chromatography; phosphopeptides are enriched using phosphospecific antibodies, immobilized metal affinity chromatography or titanium dioxide (TiO_2) chromatography; phosphopeptides are analyzed using mass spectrometry, and peptides are sequenced and analyzed.

Tools and Methods

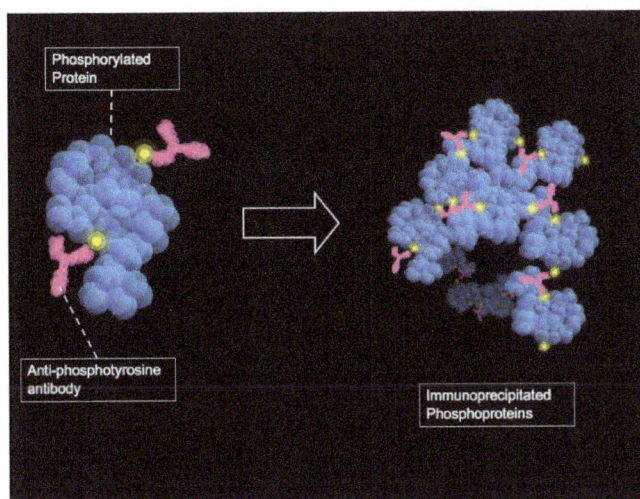

Method of phosphoprotein purification by immunoprecipitation
with anti-phosphotyrosine antibodies.

The analysis of the entire complement of phosphorylated proteins in a cell is certainly a feasible option. This is due to the optimization of enrichment protocols for phosphoproteins and phosphopeptides, better fractionation techniques using chromatography, and improvement of methods to selectively visualize phosphorylated residues using mass spectrometry. Although the current procedures for phosphoproteomic analysis are greatly improved, there is still sample loss and inconsistencies with regards to sample preparation, enrichment, and instrumentation. Bioinformatics tools and biological sequence databases are also necessary for high-throughput phosphoproteomic studies.

Enrichment Strategies

Previous procedures to isolate phosphorylated proteins included radioactive labeling with P-labeled ATP followed by SDS polyacrylamide gel electrophoresis or thin layer chromatography. These traditional methods are inefficient because it is impossible to obtain large amounts of proteins required for phosphorylation analysis. Therefore, the current and simplest methods to enrich phosphoproteins are affinity purification using phosphospecific antibodies, immobilized metal affinity chromatography (IMAC), strong cation exchange (SCX) chromatography, or titanium dioxide chromatography.

Antiphosphotyrosine antibodies have been proven very successful in purification, but fewer reports have been published using antibodies against phosphoserine- or phosphothreonine-containing proteins. IMAC enrichment is based on phosphate affinity for immobilized metal chelated to the resin. SCX separates phosphorylated from non-phosphorylated peptides based on the negatively charged phosphate group. Titanium dioxide chromatography is a newer technique that requires significantly less column preparation time. Many phosphoproteomic studies use a combination of these enrichment strategies to obtain the purest sample possible.

Mass Spectrometry Analysis

Mass spectrometry is currently the best method to adequately compare pairs of protein samples. The two main procedures to perform this task are using isotope-coded affinity tags (ICAT) and stable isotopic amino acids in cell culture (SILAC). In the ICAT procedure samples are labeled individually after isolation with mass-coded reagents that modify cysteine residues. In SILAC, cells are cultured separately in the presence of different isotopically labeled amino acids for several cell divisions allowing cellular proteins to incorporate the label. Mass spectrometry is subsequently used to identify phosphoserine, phosphothreonine, and phosphotyrosine-containing peptides.

Signal Transduction Studies

Intracellular signal transduction is primarily mediated by the reversible phosphorylation of various signalling molecules by enzymes dubbed kinases. Kinases transfer phosphate groups from ATP to specific serine, threonine or tyrosine residues of target molecules. The resultant phosphorylated protein may have altered activity level, subcellular localization or tertiary structure.

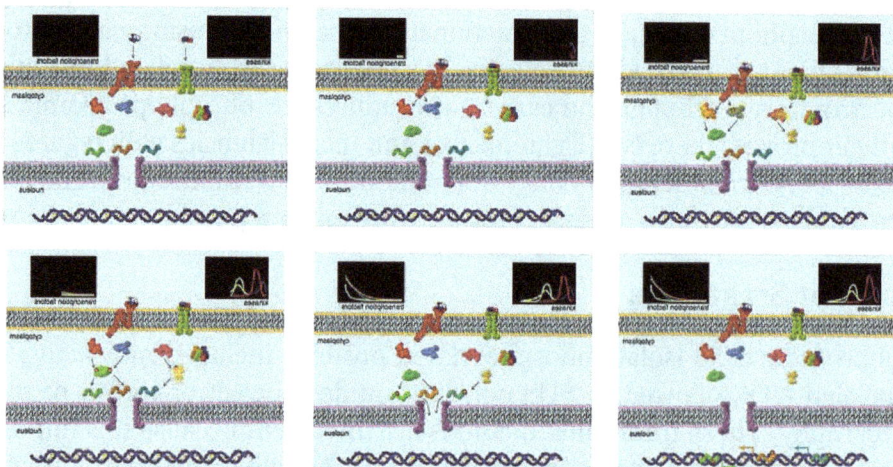

Analysis of signal transduction dynamics.

Phosphoproteomic analyses are ideal for the study of the dynamics of signalling networks. In one study design, cells are exposed to SILAC labelling and then stimulated by a specific growth factor. The cells are collected at various timepoints, and the lysates are combined for analysis by tandem MS. This allows experimenters to track the phosphorylation state of many phosphoproteins in the cell over time. The ability to measure the global phosphorylation state of many proteins at various time points makes this approach much more powerful than traditional biochemical methods for analyzing signalling network behavior.

One study was able to simultaneously measure the fold-change in phosphorylation state of 127 proteins between unstimulated and EphrinB1-stimulated cells. Of these 127 proteins, 40 showed increased phosphorylation with stimulation by EphrinB1. The researchers were able to use this information in combination with previously published data to construct a signal transduction network for the proteins downstream of the EphB2 receptor.

Another recent phosphoproteomic study included large-scale identification and quantification of phosphorylation events triggered by the anti-diuretic hormone vasopressin in kidney collecting duct. A total of 714 phosphorylation sites on 223 unique phosphoproteins were identified, including three novel phosphorylation sites in the vasopressin-sensitive water channel aquaporin-2 (AQP2).

Cancer Research

Since the inception of phosphoproteomics, cancer research has focused on changes to the phosphoproteome during tumor development. Phosphoproteins could be cancer markers useful to cancer diagnostics and therapeutics. In fact, research has shown that there are distinct phosphotyrosine proteomes of breast and liver tumors. There is also evidence of hyperphosphorylation at tyrosine residues in breast tumors but not in normal tissues. Findings like these suggest that it is possible to mine the tumor phosphoproteome for potential biomarkers.

Increasing amounts of data are available suggesting that distinctive phosphoproteins exist in various tumors and that phosphorylation profiling could be used to fingerprint cancers from different origins. In addition, systematic cataloguing of tumor-specific phosphoproteins in individual patients could reveal multiple causative players during cancer formation. By correlating this experimental data to clinical data such as drug response and disease outcome, potential cancer markers could be identified for diagnosis, prognosis, prediction of drug response, and potential drug targets.

Limitations

While phosphoproteomics has greatly expanded knowledge about the numbers and types of phosphoproteins, along with their role in signaling networks, there are still several limitations to these techniques. To begin with, isolation methods such as anti-phosphotyrosine antibodies do not distinguish between isolating tyrosine-phosphorylated proteins and proteins associated with tyrosine-phosphorylated proteins. Therefore, even though phosphorylation dependent protein-protein interactions are very important, it is important to remember that a protein detected by this method is not necessarily a direct substrate of any tyrosine kinase. Only by digesting the samples before immunoprecipitation can isolation of only phosphoproteins and temporal profiles of individual phosphorylation sites be produced. Another limitation is that some relevant proteins will likely be missed since no extraction condition is all encompassing. It is possible that proteins with low stoichiometry of phosphorylation, in very low abundance, or phosphorylated as a target for rapid degradation will be lost. Bioinformatics analyses of low-throughput phosphorylation data together with high-throughput phosphoproteomics data (based mostly on MS/MS) estimate that current high-throughput protocols, after several repetitions are capable of capturing 70% to 95% of total phosphoproteins, but only 40% to 60% of total phosphorylation sites.

IMMUNOPROTEOMICS

Immunoproteomics is the study of large sets of proteins (proteomics) involved in the immune response.

Examples of common applications of immunoproteomics include:

- The isolation and mass spectrometric identification of MHC (major histocompatibility complex) binding peptides.

- Purification and identification of protein antigens binding specific antibodies (or other affinity reagents).

- Comparative immunoproteomics to identify proteins and pathways modulated by a specific infectious organism, disease or toxin.

The identification of proteins in immunoproteomics is carried out by techniques including gel based, microarray based, and DNA based techniques, with mass spectroscopy typically being the ultimate identification method.

Applications

Immunology

Immunoproteomics is and has been used to increase scientific understanding of both autoimmune disease pathology and progression. Using biochemical techniques, gene and ultimately protein expression can be measured with high fidelity. With this information, the biochemical pathways causing pathology in conditions such as multiple sclerosis and Crohn's disease can potentially be elucidated. Serum antibody identification in particular has proven to be very useful as a diagnostic tool for a number of diseases in modern medicine, in large part due to the relatively high stability of serum antibodies.

Immunoproteomic techniques are additionally used for the isolation of antibodies. By identifying and proceeding to sequence antibodies, scientists are able to identify potential protein targets of said antibodies. In doing so, it is possible to determine the antigen(s) responsible for a particular immune response. Identification and engineering of antibodies involved in autoimmune disease pathology may offer novel techniques in disease therapy.

Drug Engineering

By identifying the antigens responsible for a particular immune response, it is possible to identify viable targets for novel drugs. In addition, specific antigens can further be classified based on immunoreactivity for identification of future potential vaccine preparations. In addition to the identification of vaccine candidates, immunoproteomic techniques such as western blotting can additionally be used for measuring the efficacy of a given vaccine.

Technology and Instrumentation

Mass Spectrometry

Mass spectrometry can be used in the sequencing of MHC binding motifs, which can subsequently be used to predict T cell epitopes. The technique of peptide mass fingerprinting (PMF) can be used to check a peptide's mass spectrum against a database of protein digests which have already been documented. If the mass spectrum of the protein of interest as well as the database protein share a large amount of homology, it is likely that the protein of interest is contained within the sample.

2-D Gel Electrophoresis and Western Blotting

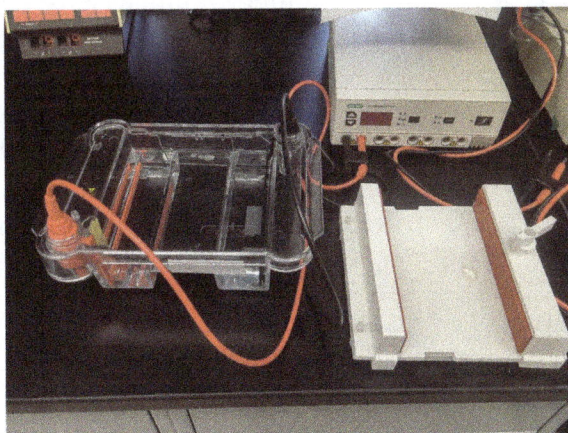

Common 2D-gel electrophoresis apparatus.

Two-dimensional gel electrophoresis (2-D gel) techniques in culmination with western blotting has been used for many years in the identification of immune response magnitude. This can be accomplished by comparing various samples against molecular-weight size markers for qualitative analysis and against known amounts of protein standards for quantitative analysis.

2-D Liquid Chromatography

By coupling liquid chromatography with a variety of other immunodetection techniques such as serological proteome analysis (SERPA), it is possible to analyze the hydrophobicity, PI, relative mass, and antibody reactivity of antibodies within a given serum.

Microarray

Microarray analysis of various serums can be used as a means to identify changes in gene expression before, after, and during a given immune response.

SECRETOMICS

Secretomics is a subset of proteomics in which all of the secreted proteins of a cell, tissue, or organism are analyzed. Secreted proteins are involved in a variety of physiological processes, including cell signaling and matrix remodeling, but are also integral to invasion and metastasis of malignant cells. Secretomics has thus been especially important in the discovery of biomarkers for cancer and understanding molecular basis of pathogenesis.

In 2000 Tjalsma et al. coined the term 'secretome' in their study of the eubacterium B. subtilis. They defined the secretome as all of the secreted proteins and secretory machinery of the bacteria. Using a database of protein sequences in B. subtilis and an algorithm that looked at cleavage sites and amino-terminal signal peptides characteristic of secreted proteins they were able to predict what fraction of the proteome is secreted by the cell. In 2001 the same lab set a standard of secretomics – predictions based on amino acid sequence alone are not enough to define the secretome. They used two-dimensional gel electrophoresis and mass spectrometry to identify 82 proteins secreted by B. subtilis, only 48 of which had been predicted using the genome-based method of their previous paper. This demonstrates the need for protein verification of predicted findings.

As the complicated nature of secretory pathways was revealed – namely that there are many non-classical pathways of secretion and there are many non-secreted proteins that are a part of the classical secretory pathway – a more in-depth definition of the secretome became necessary. In 2010, Agrawal et al. suggested defining the secretome as "the global group of secreted proteins into the extracellular space by a cell, tissue, organ, or organism at any given time and conditions through known and unknown secretory mechanisms involving constitutive and regulated secretory organelles".

Challenges of Secretomic Analysis

Contaminants

In culture, cells are surrounded by contaminants. Bovine serum from cell culture media and cellular debris can contaminate the collection of secreted proteins used for analysis. Bovine contaminants present a particular challenge because the protein sequences of many bovine extracellular proteins, like fibronectin and fibulin-1, are similar to the human protein sequences. To remove these contaminants, cells can be washed with PBS or serum-free medium (SFM) before incubating in SFM and collecting secreted proteins. Care must be taken not to burst cells, releasing intracellular proteins. In addition, incubation time and conditions must be optimized so that the metabolic stress that can be induced by the lack of nutrients in SFM does not affect secretomic analysis.

Low Concentration

Some proteins are secreted in low abundance and then diluted further in the cell culture medium or body fluid, making these proteins difficult to detect and analyze. Concentration methods like TCA precipitation can be used as well as highly sensitive methods like antibody microarrays that can detect even single molecules of a protein.

Relevance of in Vitro Studies

Many secretomic studies are conducted in vitro with cell culture methods, but it is unclear whether the same proteins are secreted in vivo. More and more studies, especially those looking at the cancer secretome, are using in vivo methods to confirm the relevance of the results obtained in vitro. For example, proximal biological fluids can be collected adjacent to a tumor in order to conduct a secretomic analysis.

Methods

Genome-wide Prediction

Many secreted proteins have an N-terminal peptide sequence that signals for the translated protein to move into the endoplasmic reticulum where the processing occurs that will ultimately lead to secretion. The presence of these signal peptides can be used to predict the secretome of a cell. Software such as SignalP can identify signal sequences (and their cleavage sites) to predict proteins that are secreted. Since transmembrane proteins are also processed in the ER, but not secreted, software like the TMHMM server is used to predict transmembrane domains and therefore eliminate false positives. Some secretory proteins do not have classical signal peptide sequences. These 'leaderless secretory proteins' (LSPs) will be missed by SignalP. SecretomeP is a software that has been developed to try to predict these non-classical secretory proteins from their sequences. Genome-wide secretomes have been predicted for a wide range of organisms, including human, mouse, zebrafish, and hundreds of bacteria.

Genome-wide prediction methods have a variety of problems. There is a high possibility of false positives and false negatives. In addition, gene expression is heavily influenced by environmental conditions, meaning a secretome predicted from the genome or a cDNA library is not likely to match completely with the true secretome. Proteomic approaches are necessary to validate any predicted secreted proteins.

Several genome-wide secretome databases or knowledgebases are available based on both curation and computational prediction. These databases include: the fungal secretome database (FSD), the fungal secretome knowledgebase (FunSecKB), (FunSecKB2), (PlantSecKB), and the lactic acid bacterial secretome database. The human and animal protein subcellular location database (MetaSecKB) and the protist subcellular proteome database (ProtSecKB) are also recently released. Though there are some inaccuracies in the computational prediction, these databases provide useful resources for further characterizing the protein subcellular locations.

Proteomic Approaches

Mass spectrometry analysis is integral to secretomics. Serum or supernatant containing secreted proteins is digested with a protease and the proteins are separated by 2D gel electrophoresis or chromatographic methods. Each individual protein is then analyzed by mass spectrometry and the peptide-mass fingerprint generated can be run through a database to identify the protein.

Stable isotope labeling by amino acids in cell culture (SILAC) has emerged as an important method in secretomics – it helps to distinguish between secreted proteins and bovine serum contaminants in cell culture. Supernatant from cells grown in normal medium and cells grown in medium with stable-isotope labeled amino acids is mixed in a 1:1 ratio and analyzed by mass spectrometry. Protein contaminants in the serum will only show one peak because they do not have a labeled equivalent. As an example, the SILAC method has been used successfully to distinguish between proteins secreted by human chondrocytes in culture and serum contaminants.

An antibody microarray is a highly sensitive and high-throughput method for protein detection that has recently become part of secretomic analysis. Antibodies, or another type of binder molecule, are fixed onto a solid support and a fluorescently labeled protein mixture is added. Signal intensities are used to identify proteins. Antibody microarrays are extremely versatile – they can be used to analyze the amount of protein in a mixture, different protein isoforms, posttranslational modifications, and the biochemical activity of proteins. In addition, these microarrays are highly sensitive – they can detect single molecules of protein. Antibody microarrays are currently being used mostly to analyze human plasma samples but can also be used for cultured cells and body fluid secretomics, presenting a simple way to look for the presence of many proteins at one time.

Implications and Significance

Discovery of Cancer Biomarkers

Besides being important in normal physiological processes, secreted proteins also have an integral role in tumorigenesis through cell growth, migration, invasion, and angiogenesis, making secretomics an excellent method for the discovery of cancer biomarkers. Using a body fluid or full serum proteomic method to identify biomarkers can be extremely difficult – body fluids are complex and highly variable. Secretomic analysis of cancer cell lines or diseased tissue presents a simpler and more specific alternative for biomarker discovery.

The two main biological sources for cancer secretomics are cancer cell line supernatants and proximal biological fluids, the fluids in contact with a tumor. Cancer cell line supernatant is an attractive source of secreted proteins. There are many standardized cell lines available and supernatant is much simpler to analyze than proximal body fluid. But it is unclear whether a cell line secretome is a good representation of an actual tumor in its specific microenvironment and a standardized cell line is not illustrative of the heterogeneity of a real tumor. Analysis of proximal fluids can give a better idea of a human tumor secretome, but this method also has its drawbacks. Procedures for collecting proximal fluids still need to be standardized and non-malignant controls are needed. In addition, environmental and genetic differences between patients can complicate analysis.

Secretomic analysis has discovered potential new biomarkers in many cancer types, including lung cancer, liver cancer, pancreatic cancer, colorectal cancer, prostate cancer, and breast cancer. Prostate-specific antigen (PSA), the current standard biomarker for prostate cancer, has a low diagnostic specificity – PSA levels can not always discriminate between aggressive and non-aggressive cancer – and so a better biomarker is greatly needed. Using secretomic analysis of prostate cell lines, one study was able to discover multiple proteins found in higher levels in the serum of cancer patients than in healthy controls.

There is also a great need for biomarkers for the detection of breast cancer – currently biomarkers only exist for monitoring later stages of cancer. Secretomic analysis of breast cancer cell lines led to the discovery of the protein ALCAM as a new biomarker with promising diagnostic potential.

Assisted Reproductive Technologies

Analyzing the human embryonic secretome could be helpful in finding a non-invasive method for determining viability of embryos. In IVF, embryos are assessed on morphological criteria in an attempt to find those with high implantation potential. Finding a more quantitative method of assessment could help reduce the number of embryos used in IVF, thereby reducing higher order pregnancies. For example, one study was able to develop secretome fingerprints for many blastocysts and found 9 proteins that could distinguish between blastocysts with normal and abnormal numbers of chromosomes.

This type of analysis could help replace preimplantation genetic screening (PGS), which involves biopsy of embryonic cells and can be harmful to development.

ACTIVITY-BASED PROTEOMICS

Activity-based proteomics, or activity-based protein profiling (ABPP) is a functional proteomic technology that uses chemical probes that react with mechanistically related classes of enzymes.

The basic unit of ABPP is the probe, which typically consists of two elements: a reactive group (RG, sometimes called a "warhead") and a tag. Additionally, some probes may contain a binding group which enhances selectivity. The reactive group usually contains a specially designed electrophile that becomes covalently-linked to a nucleophilic residue in the active site of an active enzyme. An enzyme that is inhibited or post-translationally modified will not react with an activity-based probe. The tag may be either a reporter such as a fluorophore or an affinity label such as biotin or an alkyne or azide for use with the Huisgen 1,3-dipolar cycloaddition (also known as click chemistry).

Advantages

A major advantage of ABPP is the ability to monitor the availability of the enzyme active site directly, rather than being limited to protein or mRNA abundance. With classes of enzymes such as the serine hydrolases and metalloproteases that often interact with endogenous inhibitors or that exist as inactive zymogens, this technique offers a valuable advantage over traditional techniques that rely on abundance rather than activity.

Multidimensional Protein Identification Technology

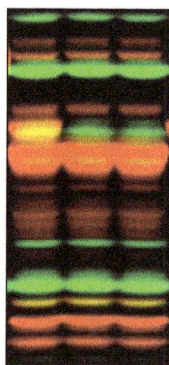

In-gel ABPP using probes with different fluorophores in the same lane to simultaneously profile differences in enzyme activities.

In recent years ABPP has been combined with tandem mass spectrometry enabling the identification of hundreds of active enzymes from a single sample. This technique, known as *ABPP-MudPIT* (multidimensional protein identification technology) is especially useful for profiling inhibitor selectivity as the potency of an inhibitor can be tested against hundreds of targets simultaneously.

NEUROPROTEOMICS

Neuroproteomics is the study of the protein complexes and species that make up the nervous system. These proteins interact to make the neurons connect in such a way to create the intricacies that nervous system is known for. Neuroproteomics is a complex field that has a long way to go in terms of profiling the entire neuronal proteome. It is a relatively recent field that has many applications in therapy and science. So far, only small subsets of the neuronal proteome have been mapped, and then only when applied to the proteins involved in the synapse.

The word proteomics was first used in 1994 by Marc Wilkins as the study of "the protein equivalent of a genome". It is defined as all of the proteins expressed in a biological system under specific physiologic conditions at a certain point in time. It can change with any biochemical alteration, and so it can only be defined under certain conditions. Neuroproteomics is a subset of this field dealing with the complexities and multi-system origin of neurological disease. Neurological function is based on the interactions of many proteins of different origin, and so requires a systematic study of subsystems within its proteomic structure.

Modern Times

Neuroproteomics has the difficult task of defining on a molecular level the pathways of consciousness, senses, and self. Neurological disorders are unique in that they do not always exhibit outward symptoms. Defining the disorders becomes difficult and so neuroproteomics is a step in the right direction of identifying bio-markers that can be used to detect diseases. Not only does the field have to map out the different proteins possible from the genome, but there are many modifications that happen after transcription that affect function as well. Because neurons are such dynamic structures, changing with every action potential that travels through them, neuroproteomics offers the most potential for mapping out the molecular template of their function. Genomics offers a static roadmap of the cell, while proteomics can offer a glimpse into structures smaller than the cell because of its specific nature to each moment in time.

Mechanisms of Use

Protein Separation

In order for neuroproteomics to function correctly, proteins must be separated in terms of the proteome from which they came. For example, one set might be under normal conditions, while another might be under diseased conditions. Proteins are commonly separated using two-dimensional polyacrylamide gel electrophoresis (2D PAGE). For this technique, proteins are run across an immobile gel with a pH gradient until they stop at the point where their net charge is neutral. After separating by charge in one direction, sodium dodecyl sulfate is run in the other direction to separate the proteins by size. A two-dimensional map is created using this technique that can be used to match additional proteins later. One can usually match the function of a protein by identifying in an 2D PAGE in simple proteomics because many intracellular somatic pathways are known. In neuroproteomics, however, many proteins combine to give an end result that may be neurological disease or breakdown. It is necessary then to study each protein individually and find a correlation between the different proteins to determine the cause of a neurological disease. New techniques are being developed that can identify proteins once they are separated out using 2D PAGE.

Protein Identification

Protein separate techniques, such as 2D PAGE, are limited in that they cannot handle very high or low molecular weight protein species. Alternative methods have been developed to deal with such cases. These include liquid chromatography mass spectrometry along with sodium dodecyl sulfate polyacrylamide gel electrophoresis, or liquid chromatography mass spectrometry run in multiple dimensions. Compared to simple 2D page, liquid chromatography mass spectrometry can handle a larger range of protein species size, but it is limited in the amount of protein sample it handle at once. Liquid chromatography mass spectrometry is also limited in its lack of a reference map from which to work with. Complex algorithms are usually used to analyze the fringe results that occur after a procedure is run. The unknown portions of the protein species are usually not analyzed in favor of familiar proteomes, however. This fact reveals a fault with current technology; new techniques are needed to increase both the specificity and scope of proteome mapping.

Applications

Drug Addiction

It is commonly known that drug addiction involves permanent synaptic plasticity of various neuronal circuits. Neuroproteomics is being applied to study the effect of drug addiction across the synapse. Research is being conducted by isolating distinct regions of the brain in which synaptic transmission takes place and defining the proteome for that particular region. Different stages of drug abuse must be studied, however, in order to map out the progression of protein changes along the course of the drug addiction. These stages include enticement, ingesting, withdrawal, addiction, and removal. It begins with the change in the genome through transcription that occurs due to the abuse of drugs. It continues to identify the most likely proteins to be affected by the drugs and focusing in on that area. For drug addiction, the synapse is the most likely target as it involves communication between neurons. Lack of sensory communication in neurons is often an outward sign of drug abuse, and so neuroproteomics is being applied to find out what proteins are being affected to prevent the transport of neurotransmitters. In particular, the vesicle releasing process is being studied to identify the proteins involved in the synapse during drug abuse. Proteins such as synaptotagmin and synaptobrevin interact to fuse the vesicle into the membrane. Phosphorylation also has its own set of proteins involved that work together to allow the synapse to function properly. Drugs such as morphine change properties such as cell adhesion, neurotransmitter volume, and synaptic traffic. After significant morphine application, tyrosine kinases received less phosphorylation and thus send fewer signals inside the cell. These receptor proteins are unable to initiate the intracellular signaling processes that enable the neuron to live, and necrosis or apoptosis may be the result. With more and more neurons affected along this chain of cell death, permanent loss of sensory or motor function may be the result. By identifying the proteins that are changed with drug abuse, neuroproteomics may give clinicians even earlier biomarkers to test for to prevent permanent neurological damage.

Recently, a novel terminology (Psychoproteomics) has been coined by the University of Florida researchers from Dr. Mark S Gold Lab. Kobeissy et al. defined Psychoproteomics as integral proteomics approach dedicated to studying proteomic changes in the field of psychiatric disorders, particularly substance-and-drug-abuse neurotoxicity.

Brain Injury

Traumatic brain injury is defined as a "direct physical impact or trauma to the head followed by a dynamic series of injury and repair events". Recently, neuroproteomics have been applied to studying the disability that over 5.4 million Americans live with. In addition to physically injuring the brain tissue, traumatic brain injury induces the release of glutamate that interacts with ionotropic glutamate receptors (iGluRs). These glutamate receptors acidify the surrounding intracranial fluid, causing further injury on the molecular level to nearby neurons. The death of the surrounding neurons is induced through normal apoptosis mechanisms, and it is this cycle that is being studied with neuroproteomics. Three different cysteine protease derivatives are involved in the apoptotic pathway induced by the acidic environment triggered by glutamate. These cysteine proteases include calpain, caspase, and cathepsin. These three proteins are examples of detectable signs of traumatic brain injury that are much more specific than temperature, oxygen level, or intracranial pressure. Proteomics thus also offers a tracking mechanism by which researchers can monitor the progression of traumatic brain injury, or a chronic disease such as Alzheimer's or Parkinson's. Especially in Parkinson's, in which neurotransmitters play a large role, recent proteomic research has involved the study of synaptotagmin. Synaptotagmin is involved in the calcium-induced budding of vesicle containing neurotransmitters from the presynaptic membrane. By studying the intracellular mechanisms involved in neural apoptosis after traumatic brain injury, researchers can create a map that genetic changes can follow later on.

Nerve Growth

One group of researchers applied the field of neuroproteomics to examine how different proteins affect the initial growth of neuritis. The experiment compared the protein activity of control neurons with the activity of neurons treated with nerve growth factor (NGF) and JNJ460, an "immunophilin ligand." JNJ460 is an offspring of another drug that is used to prevent immune attack when organs are transplanted. It is not an immunosuppressant, however, but rather it acts as a shield against microglia. NGF promotes neuron viability and differentiation by binding to TrkA, a tyrosine receptor kinase. This receptor is important in initiating intracellular metabolic pathways, including Ras, Rak, and MAP kinase.

Protein differentiation was measured in each cell sample with and without treatment by NGF and JNJ460. A peptide mixture was made by washing off unbound portions of the amino acid sequence in a reverse column. The resulting mixture was then suspended a peptide mixture in a bath of cation exchange fluid. The proteins were identified by splicing them with trypsin and then searching through the results of passing the product through a mass spectrometer. This applies a form of liquid chromatography mass spectrometry to identify proteins in the mixture JNJ460 treatment resulted in an increase in "signal transduction" proteins, while NGF resulted in an increase in proteins associated with the ribosome and synthesis of other proteins. JNJ460 also resulted in more structural proteins associated with intercellular growth, such as actin, myosin, and troponin. With NGF treatment, cells increased protein synthesis and creation of ribosomes. This method allows the analysis of all of the protein patterns overall, rather than a single change in an amino acid. Western blots confirmed the results, according to the researchers, though the changes in proteins were not as obvious in their protocol.

The main significance to these findings are that JNJ460 are NGF are distinct processes that both control the protein output of the cell. JNJ460 resulted in increased neuronal size and stability while NGF resulted in increased membrane proteins. When combined together, they significantly increase a neuron's chance of growth. While JNJ460 may "prime" some parts of the cell for NGF treatment, they do not work together. JNJ460 is thought to interact with Schwann cells in regenerating actin and myosin, which are key players in axonal growth. NGF helps the neuron grow as a whole. These two proteins do not play a part in communication with other neurons, however. They merely increase the size of the membrane down which a signal can be sent. Other neurotrophic factor proteomes are needed to guide neurons to each other to create synapses.

Limitations

The broad scope of the available raw neuronal proteins to map requires that initial studies be focused on small areas of the neurons. When taking samples, there are a few places that interest neurologists most. The most important place to start for neurologists is the plasma membrane. This is where most of the communication between neurons takes place. The proteins being mapped here include ion channels, neurotransmitter receptors, and molecule transporters. Along the plasma membrane, the proteins involved in creating cholesterol-rich lipid rafts are being studied because they have been shown to be crucial for glutamate uptake during the initial stages of neuron formation. As mentioned before, vesicle proteins are also being studied closely because they are involved in disease. Collecting samples to study, however, requires special consideration to ensure that the reproducibility of the samples is not compromised. When taking a global sample of one area of the brain for example, proteins that are ubiquitous and relatively unimportant show up very clear in the SDS PAGE. Other unexplored, more specific proteins barely show up and are therefore ignored. It is usually necessary to divide up the plasma membrane proteome, for example, into subproteomes characterized by specific function. This allows these more specific classes of peptides to show up more clearly. In a way, dividing into subproteomes is simply applying a magnifying lens to a specific section of a global proteome's SDS PAGE map. This method seems to be most effective when applied to each cellular organelle separately. Mitochondrial proteins, for example, which are more effective at transporting electrons across its membrane, can be specifically targeted effectively in order to match their electron-transporting ability to their amino acid sequence.

QUANTITATIVE PROTEOMICS

Quantitative proteomics is a powerful approach used for both discovery and targeted proteomic analyses to understand global proteomic dynamics in a cell, tissue or organism. Most quantitative proteomic analyses entail the isotopic labeling of proteins or peptides in the experimental groups, which can then be differentiated by mass spectrometry. Relative quantitation methods (SILAC, ICAT, ICPL and isobaric tags) are used to compare protein or peptide abundance between samples, while spiking unlabeled samples with known concentrations of isotopically-labeled synthetic peptides can yield absolute quantitation of target peptides via selected reaction monitoring (SRM). Label-free strategies are also available for both relative and absolute quantitation. Although these strategies are more complex than mere protein identification, quantitative proteomics is critical for our understanding of global protein expression and modifications underlying the molecular mechanisms of biological processes and disease states.

Early biochemical proteomics research focused on identifying and understanding the functions of individual proteins or protein complexes. Technological advances in instrumentation, however, have increased the number of proteins that one can analyze in a single sample from hundreds a decade ago to thousands today. At this level of analysis, global protein dynamics can be studied on a cellular, tissue or even organismal level. This type of approach is consistent with the increasingly broad-scope analyses that are being used in other life science fields, including genomics, transcriptomics, metabolomics and kinomics, which are giving us a greater understanding of global biological processes and how they respond to different stimuli or change during disease states.

While proteomic analyses can be used to qualitatively identify thousands of proteins in cells or other biological samples, there is also a need to quantitate these proteins. Because of the dynamic and interactive nature of proteins, quantitative proteomics is considerably more complex than simply identifying proteins in a sample. Due of the considerable amount of data that one can acquire from quantitative proteomics, this approach is critical for our understanding of global protein kinetics and molecular mechanisms of biological processes.

Two fundamental approaches to proteomic analyses are currently employed. In top-down proteomics, intact proteins or large fragments are ionized and analyzed by mass spectrometry (MS). Bottom-up proteomics analysis relies on peptides, which are generated by proteolytic digestion of protein samples. Due to the protein size limitation in top-down proteomics (<50 kD), bottom-up proteomics analysis is more commonly used.

Because of the overwhelming number of proteotypic peptides in a sample, only a small subset of all peptides can be analyzed in a single MS run, limiting the number of proteins that are identified. The number of proteins available for quantitation is limited even further because they must be identified in all samples that are tested in a single experiment. Practically speaking, the linear dynamic range of quantitation is often limited by 10- to 20-fold depending on the sensitivity of the instrument and complexity of the sample. This limitation affects the scope of quantitative proteomics.

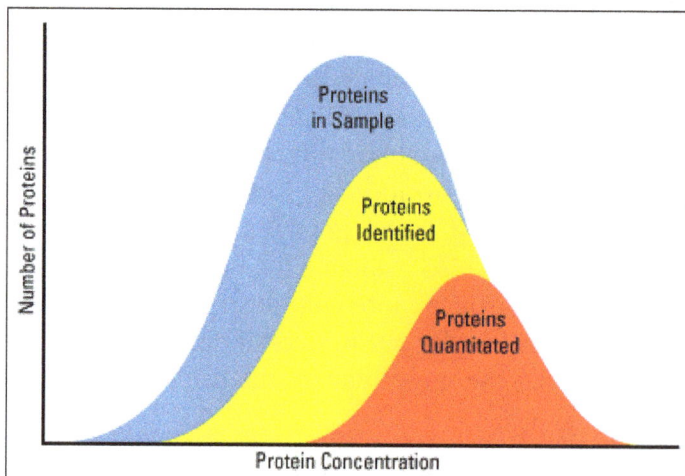

Protein abundance and sample complexity. Protein abundance and sample complexity are significant factors that affect the availability of proteins for mass spectrometric quantitation.

Sample complexity is a critical factor for peptide quantitation because identification and quantification rates are directly proportional to sample complexity. Methods such as affinity purification are often performed to remove high-abundance proteins and reduce sample complexity. In-line liquid chromatography (LC) is also a common pre-MS fractionation process that chemically separates peptides and further reduces sample complexity.

Quantitative proteomic analyses typically rely on MS to identify or quantitate selected peptides, although tandem mass spectrometry (MS/MS) is required for peptide identification. During the first round of MS (MS1), ionized peptides are sampled to produce a precursor ion spectrum that represents all ionized peptides in the sample. Individual ions are then selected to undergo collision-induced fragmentation (CID) and a second round of MS (MS2), which yields a fragment ion spectrum for each precursor ion. These fragment spectra are compared to peptide databases and assigned specific peptide sequences, then computationally organized into the predicted protein sequence.

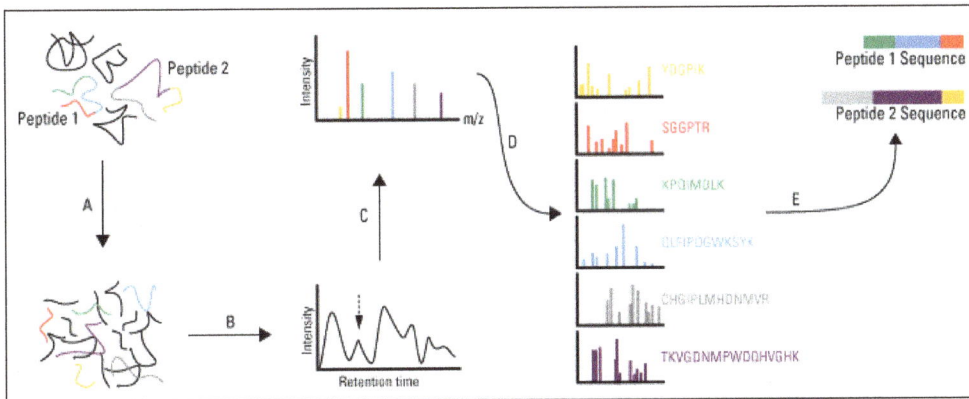

Overview of proteomic analysis by MS/MS. Sample proteins are extracted and digested into peptides (A). The sample complexity may then be reduced prior to chemical separation by LC (B). Fractions (indicated by dotted arrow) are then analyzed by MS (C), during which the peptides are ionized and their mass-to-charge ratio (m/z) measured to yield a precursor ion spectrum. Selected ions are then fragmented by collision-induced dissociation (CID) and the individual fragment ions measured by MS (D). The fragment ion spectra are then assigned peptide sequences based on database comparison and protein sequences are predicted (E).

Discovery vs. Targeted Proteomics

Strategies to improve the sensitivity and scope of proteomic analysis generally require large sample quantities and multi-dimensional fractionation, which sacrifices throughput. Alternatively, efforts to improve the sensitivity and throughput of protein quantification limit the number of features that can be monitored.

For this reason, proteomics research is typically divided into two categories: discovery proteomics and targeted proteomics. Discovery proteomics optimizes protein identification by spending more time and effort per sample and reducing the number of samples analyzed. In contrast, targeted proteomics strategies limit the number of features that will be monitored and then optimize the chromatography, instrument tuning, and acquisition methods to achieve the highest sensitivity and throughput for hundreds or thousands of samples.

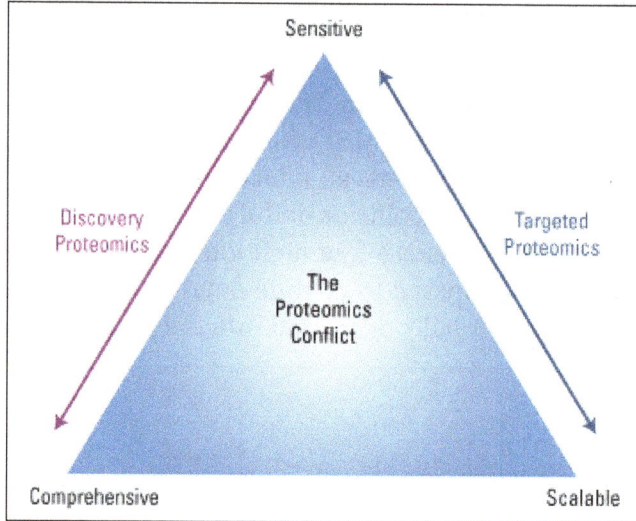

The balance between scope, sensitivity and scalability of discovery and targeted proteomics. Due to the broad-scope nature and sensitivity of discovery proteomics, the ability to perform a comprehensive analysis of hundreds or thousands of samples is limited. Conversely, targeted proteomic analysis entails the quantitation of discrete subsets of peptides, which allows researchers to analyze these peptides across thousands of samples with the highest level of sensitivity.

Discovery proteomics experiments are intended to identify as many proteins as possible across a broad dynamic range and often require depletion of highly abundant proteins, enrichment of relevant components (e.g., subcellular compartments or protein complexes), and fractionation to decrease sample complexity (e.g., SDS-PAGE or chromatography). These strategies can reduce the dynamic range between components in a fraction and reduce the competition between proteins or peptides for ionization and MS duty cycle time. Quantitative discovery proteomics experiments add a further challenge because they seek to identify and quantify protein levels across multiple fractionated samples.

Targeted proteomics experiments are typically designed to quantify less than 100 proteins with very high precision, sensitivity, specificity and throughput. Indeed, this approach typically minimizes the amount of sample preparation to improve precision and throughput. Targeted MS quantitation strategies use specialized workflows and instruments to improve the specificity and quantification of a limited number of features across hundreds or thousands of samples, including directed sequencing by inclusion lists and selected (or multiple) reaction monitoring (SRM or MRM, respectively).

While discovery proteomics analysis is most often used to inventory proteins in a sample or detect differences in the abundance of proteins between multiple samples, targeted quantitative proteomic experiments are increasingly used in pharmaceutical and diagnostic applications to quantify proteins and metabolites in complex samples. Additionally, targeted proteomics often follows discovery proteomics to quantitate specific proteins found during discovery screening.

The characteristics of specific mass spectrometers make them more amenable to use with either discovery or targeted proteomic analysis. For example, because discovery proteomics emphasizes identification of all peptides in a limited number of samples, high-resolution instruments, including Thermo Scientific Orbitrap mass analyzers, are used to maximize the detection of peptides with minute mass-to-charge ratio (m/z) differences. Conversely, because targeted proteomics emphasizes sensitivity and throughout, instruments including triple quadrupoles and ion traps are used.

Relative vs. Absolute Quantitation

Mass spectrometry is not inherently quantitative, because proteolytic peptides show great variability in physiochemical properties; this in turn results in mass spectrometric variability between runs. Additionally, mass spectrometers only sample a small percentage of the total peptides in a sample. Therefore, various approaches have been developed to perform relative and absolute proteomic quantitation.

Relative quantitation strategies compare the levels of individual peptides in a sample to those in an identical, but experimentally-modified, sample. One approach to relative quantitation is to separately analyze samples by MS and compare their spectra to determine peptide abundance in one sample relative to another. This is performed in label-free quantitation strategies.

More costly and time-consuming approaches require internal, isotopically-labeled standards for the mass spectrometer to distinguish between identical proteins from separate samples. A typical relative quantitation experiment that uses isotopic labels entails labeling proteins or peptides from two experimental samples with isotopically-heavy and light atoms (via a labeled amino acid or cell culture component), which makes the peptides in these two samples isotopologues (identical molecules that differ only in isotope composition).

After alteration of the proteome in the experimental group through chemical treatment or genetic manipulation, equal amounts of protein from both populations are combined and analyzed by LC-MS or LC-MS/MS analysis. Because the light and heavy forms of individual peptides are chemically identical, they co-elute during LC prefractionation and are therefore detected simultaneously during MS analysis. The peak intensities of the heavy and light peptides are then compared to determine the change in abundance in one sample relative to that of the other. Methods to isotopically label proteins or peptides include metabolic labeling of live cells and enzymatic or chemical labeling of extracted proteins or peptides.

Technique	Type	Number Samples (per LC-MS run)	Precision (% CV)	Benefits	Drawbacks	Instruments
Discovery-based (untargeted quantitative analysis coupling protein identification with quantitation)						
Label-Free	Relative	1	< 30	• Applicable to any sample type • Cost-efficient sample preparation • Minimal sample handling	• Each sample runs individually (low throughput) • Requires extremely reproducible LC separations • Requires multiple technical replicates	Orbitrap
TMT	Relative	2 to 6	< 20	• Applicable to any sample type • Multiplexing increases MS throughput	• Requires extensive fractionation or long chromatographic gradients	Orbitrap with HCD Ion Trap with PQD or Trap-HCD
SILAC	Relative	2 or 3	< 20	• Least susceptible to inter-sample variations in sample handling and preparation • Multiplexing increases MS throughput	• Only readily applicable to cell cultures • Increases MS spectral complexity	Orbitrap
Targeted (analysis of predetermined peptides from discovery-based experiments or literature)						
HR/AM-SIM	Relative or Absolute	1	< 20	• Uses the same MS system as discovery quantitation • Easy method development using Pinpoint software	• Requires reproducible LC separations	Orbitrap
iSRM	Relative or Absolute	1	< 10	• Up to 15,000 SRM transitions per run • Simultaneous protein quantitation and confirmation of identity • Suitable for determining the absolute quantity of a protein in a complex biological matrix • Easy method development using Pinpoint software	• Requires reproducible LC separations	TSQ triple quadrupole

Discovery-based and targeted quantitation techniques all have specific applications and advantages

Relative and absolute quantitation strategies each have their benefits and drawbacks.

Absolute proteomic quantitation using isotopic peptides entails spiking known concentrations of synthetic, heavy isotopologues of target peptides into an experimental sample and then performing LC-MS/MS. As with relative quantitation using isotopic labels, peptides of equal chemistry co-elute and are analyzed by MS simultaneously. Unlike relative quantitation, the abundance of the target peptide in the experimental sample is compared to that of the heavy peptide and back-calculated to the initial concentration of the standard using a pre-determined standard curve to yield the absolute quantitation of the target peptide.

Use of Relative and Absolute Quantitation Strategies

It may seem that absolute quantitation is the ideal method compared to relative quantitation because absolute peptide values from different samples can also be compared against each other to determine relative protein changes. In actuality, relative proteomic quantitation is used more often than absolute quantitation because costly reagents and time-consuming assay development are required for the absolute quantitation of each protein of interest.

Experimental bias can influence the decision to use relative or absolute quantitation strategies. One source of bias is the mass spectrometer itself, which has a limited capacity to detect low-abundance peptides in samples with a high dynamic range. Additionally, the limited duty cycle of mass spectrometers restricts the number of collisions per unit of time, which may result in an undersampling of complex proteomic samples.

Another source of bias is variation in sample preparation between experiments or individual samples in single experiments. The greater the number of steps between labeling and sample combination, the greater is the risk of introducing experimental bias. For example, during metabolic labeling, proteins are labeled in live animals or cells and the samples are then immediately combined. Because all subsequent sample preparation and analysis is performed with the combined samples, metabolic labeling has the lowest risk of experimental variation. Conversely, samples that are individually processed and analyzed in label-free quantitation strategies have a greater risk of sample variation and experimental bias.

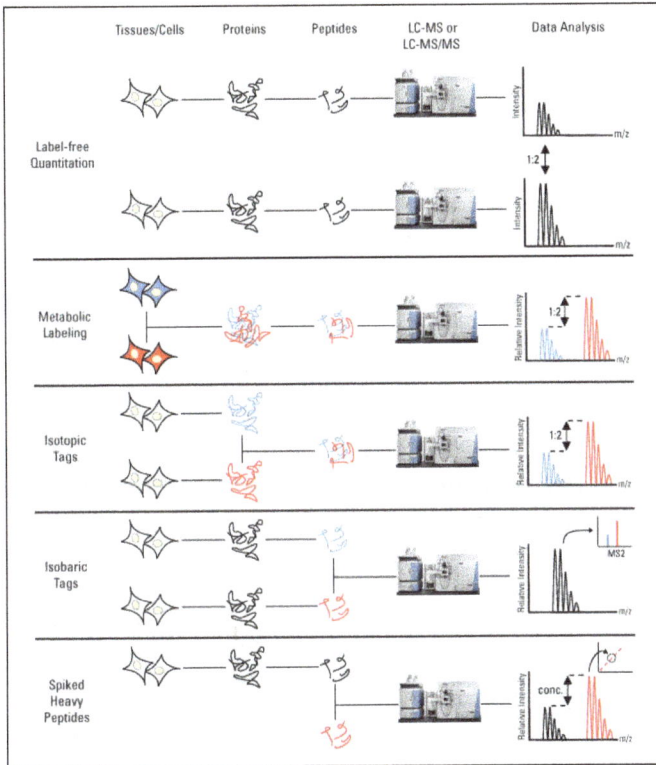

Overview of quantitative proteomics workflows. This graphic indicates the point in each workflow when samples are isotopically labeled (indicated by blue [light] and red [heavy]) for LC-MS analysis. The exception is label-free quantitation, which entails individually analyzing samples and comparing the data using multiple approaches (spectral counting and peak intensity). Metabolic labeling is characterized by the isotopic labeling of proteins in vivo, after which the samples are combined and processed for quantitative analysis. With both isotopic and isobaric tags, protein extraction occurs prior to labeling. With isobaric tags, though, LC- MS/MS analysis yields peptide fragment ion spectra generated in MS1 and the cleaved tag spectra generated in MS2, which are used for peptide identification and relative quantitation, respectively. Known quantities of heavy peptides are also spiked into unlabeled samples, and absolute quantitation is performed using a heavy peptide standard curve. Because samples are labeled and combined the earliest in the metabolic labeling workflow, this approach has the least risk of experimental bias. Conversely, label-free workflows must be tightly controlled to avoid bias, because unlabeled samples are individually analyzed.

Label-free Quantitation

Label-free methods for both relative and absolute quantitation have been developed as a rapid and low-cost alternative to other quantitative proteomic approaches. These strategies are ideal for large-sample analyses in clinical screening or biomarker discovery experiments. However, while they are good at measuring large changes in protein expression, they are less reliable for measuring small changes and can have a limited range of linear quantitative measurement (<2 orders of magnitude).

Unlike other quantitation methods, label-free samples are separately collected, prepared and analyzed by LC-MS or LC-MS/MS. Because of this, label-free quantitation experiments need to be more carefully controlled than stable isotope methods to account for any experimental variations. Protein quantitation is performed using either ion peak intensity or spectral counting.

Relative quantitation by ion peak intensity relies on LC-MS only (no MS/MS). The direct MS m/z values for all ions are detected and their signal intensities at a particular time recorded. The signal intensity from electrospray ionization has been reported to highly correlate with ion concentration, and therefore the relative peptide levels between samples can be determined directly from these peak intensities. Because of the large amount of data collected from these experiments, sensitive computer algorithms are required for automated ion peak alignment and comparison.

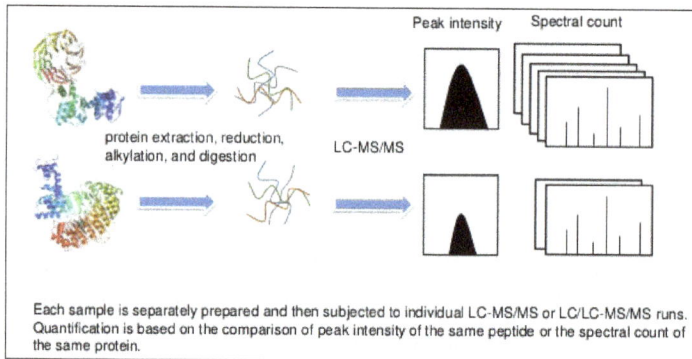

Each sample is separately prepared and then subjected to individual LC-MS/MS or LC/LC-MS/MS runs. Quantification is based on the comparison of peak intensity of the same peptide or the spectral count of the same protein.

Label-free protein quantitation methods.

Label-free protein quantitation methods are useful for measuring large changes in protein expression and can be performed rapidly and cost-effectively.

Label-free relative quantitation by spectral counts entails comparing the sum of the MS/MS spectra from a given peptide across multiple samples, which has been shown to directly correlate with protein abundance. Unlike quantitation by peak intensity, spectral counting does not require special algorithms or other tools, although significant normalization is a necessity.

Besides relative quantitation, label-free methods can be used to determine the absolute concentration of proteins in a sample. One method entails determining the exponentially modified protein abundance index (emPAI), which estimates protein abundance based on the number of peptides detected and the number of theoretically observed tryptic peptides for each protein, and which is used to determine the approximate absolute protein abundance in large-scale proteomic analyses. Another method, absolute protein expression (APEX), is based on spectral counts and uses correction factors to make protein abundance proportional to the number of peptides observed.

Metabolic Labeling

There are multiple methods of this type of in vivo labeling, and selection criteria include the extent of labeling required. Metabolic labeling for relative proteomic quantitation was first reported by Oda et al., who uniformly labeled all amino acids in yeast with heavy nitrogen (^{15}N) by growing yeast in culture medium where the only nitrogen source was ^{15}N-labeled ammonium persulfate.

This approach was further developed for use in mammalian cell lines by Mann et al., who reported a method for stable isotope labeling by amino acids in cell culture (SILAC), which has become the most common approach for in vivo isotopic labeling. Instead of labeling all amino acids with heavy nitrogen, cells are cultured in growth medium that contains $^{13}C_6$ lysine and/or $^{13}C_6$ arginine. These amino acids were chosen because trypsin, the predominant enzyme used to generate proteotypic peptides for MS analysis, cleaves at the C-terminus of lysine and arginine. Thus, all tryptic peptides from cultures grown in SILAC media (except for the very C-terminal peptides) have at least one labeled amino acid, which results in a constant mass increment in labeled samples over non-labeled, yet otherwise identical, samples.

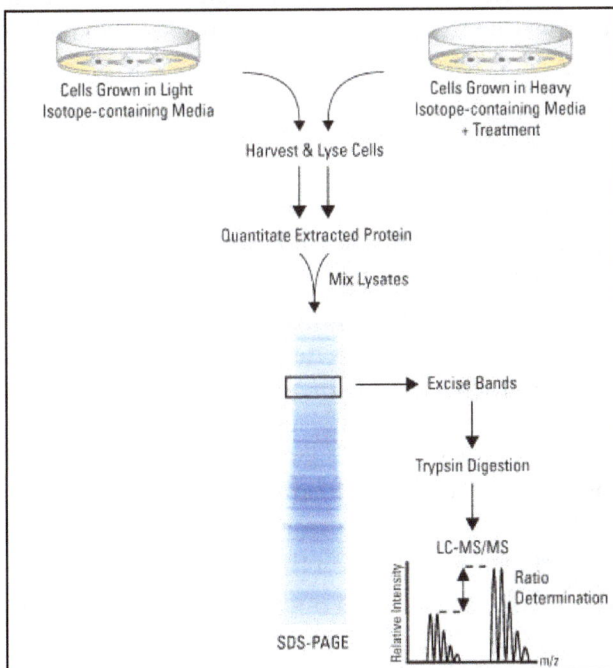

SILAC workflow. SILAC involves labeling protein samples by growing cells in media containing an isotopically heavy form of an amino acid and the naturally occurring light form. The cell lysates are then mixed, extracted and digested. When analyzed by mass spectrometry, protein level differences and posttranslational changes are easily detected.

There are many benefits to using metabolic labeling strategies compared to other methods of quantitation. For one, proteins can often attain >90% isotopic incorporation in immortalized cell lines after 6 to 8 passages. Because heavy and light samples are combined before sample preparation for MS analysis, the level of quantitation bias from processing errors is low. This key aspect of metabolic labeling makes this method particularly useful to detect relatively small changes in protein levels or posttranslational modifications between experimental conditions.

A limitation of this approach is that some cells convert high concentrations of arginine to proline, which in the case of heavy arginine labeling produces two distinct heavy peak clusters that represent heavy arginine- or proline-labeled peptides. This issue can be addressed by either accounting for the heavy proline in the quantitation calculation or by titrating the heavy arginine concentration in the culture medium to below the threshold at which conversion is detectable.

Metabolic labeling may not be amenable to cell lines that are difficult to grow or show extreme sensitivity to changes in culture medium composition. This technique also may influence how the organism functions, as growth conditions are changed to allow incorporation of heavy compounds. Finally, the number of experimental conditions per experiment is restricted when using metabolic labeling because of the limited number of heavy isotopes incorporated into lysine and arginine. For example, a maximum of three conditions per experiment (unlabeled, $^{13}C_6$- and $^{15}N_4$- labeled amino acids) can be performed with SILAC.

Isotopic Tags

For samples that are not amenable to metabolic labeling, such as when analyzing clinical samples (e.g., biological fluids, tissue samples) or when experimental time is limited, chemical or enzymatic stable isotopic labeling methods are available for quantitative proteomic analyses. These include strategies to add isotopic atoms or isotope-coded tags to peptides or proteins. While the methods described below do not comprise an exhaustive list of isotopic labeling methods, they do represent commonly used approaches.

Enzymatic labeling with ^{18}O takes advantage of the proteolytic mechanism of trypsin to incorporate two heavy oxygen atoms from $H_2^{18}O$ at the C-terminus of every newly digested peptide. In this labeling scheme, one sample is digested with trypsin and ^{18}O water and another with ^{16}O water, and then the samples are combined for relative proteomic analysis by MS. While this method is simple to execute, a disadvantage is a slow back exchange of ^{18}O and ^{16}O when the two samples are combined, leading to incomplete labeling or peptides labeled with only one heavy oxygen atom. While adding 1-5%

formic acid can attenuate this back exchange for up to 24 hours, samples labeled with this method should be processed rapidly.

Another enzymatic isotopic labeling strategy is global internal standard technology (GIST), which uses deuterated (^{2}H) acylating agents such as N-acetoxysuccinimide (NAS) to label primary amino groups on digested peptides. Acylation of these groups, though, changes the ionic states of peptides and may affect the ionization efficiency of peptides with C-terminal lysines. Additionally, isotopic methods that label with deuterium result in partial separation of heavy and light peptides during LC, because the deuterium slightly interacts with the stationary phase (e.g., C18). This difference can affect the confidence and accuracy of the internal standards, because one of them may co-elute with another peptide that inhibits its ionization.

A rapid and relatively inexpensive method of chemical labeling is stable isotope dimethylation. This approach uses formaldehyde in deuterated water to label primary amines with deuterated methyl groups. Unlike GIST, this approach does not change the ionic state of the labeled peptides because of the reductive amination that occurs, so their chemical properties remain the same as those of unlabeled peptides.

A benefit of this approach is that a wide array of sample types is amenable to formaldehyde fixation, which is fast and cheap compared to other labeling reagents. As with other methods of labeling, this method has global labeling characteristics, which has both pros and cons. While this high level of isotopic labeling is beneficial when other labeling strategies fail, it requires either using relative pure samples or sample preparation to reduce the complexity of biological samples to minimize the number of peaks detected by MS.

Commercially isotopic labeling reagents are also available that encompass a wide range of reactive groups for different crosslinker specificity and heavy labels for different applications of isotopologue separation.

Isotope-coded Affinity Tag

The isotope-coded affinity tag (ICAT) method was developed to reduce the sample complexity and identify low-abundance proteins and peptides in complex samples. ICAT tags were originally comprised of a sulfhydryl-reactive chemical crosslinking group, an 8-fold deuterated (d8; adds 8 Da to the molecular mass of the unlabeled peptide) or light (d0) linker region and a biotin molecule.

Due to the sulfhydryl-reactive chemical group, only free thiols on cysteine residues are labeled with this tag. The sample is then passed over immobilized avidin, which binds to the biotin tag and purifies the labeled peptides from the sample. Not all peptides have cysteine residues, so this method does not result in global labeling and is therefore only an inherent approach to reducing sample complexity. Once peptides are labeled, they are eluted from the sample by column chromatography using immobilized avidin or streptavidin. After purification, heavy (d8) and light (d0) samples are combined and analyzed for relative quantitation by LC-MS.

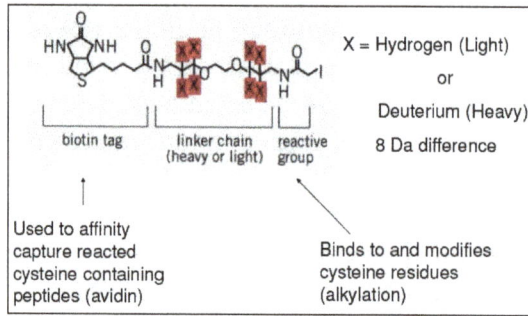

Isotope-coded affinity tag (ICAT) chemistry.

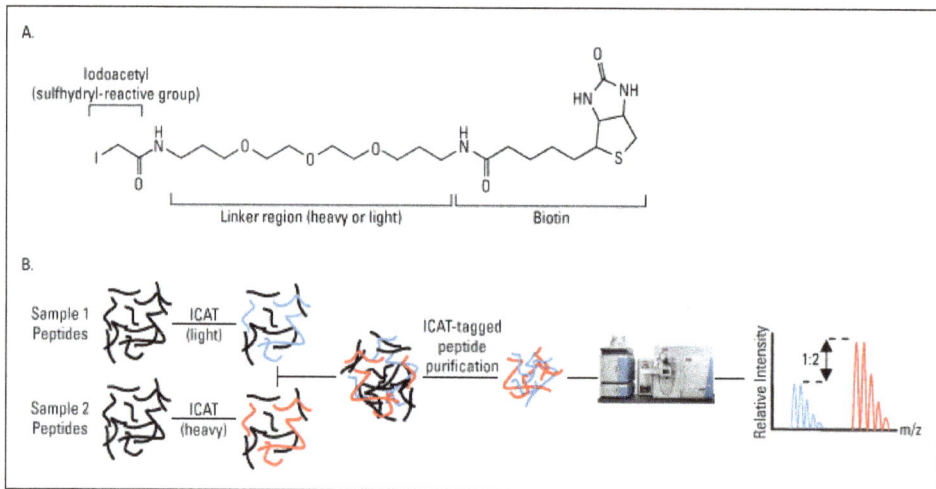

Overview of ICAT labeling and quantitation. (A) Tags consist of a sulfhydryl-reactive moiety connected to a linker region with deuterium or ^{13}C substitutions to make the tag heavy. Biotin is connected to the linker region to allow affinity purification of labeled peptides. (B) ICAT-tagged peptide purification prior to LC-MS reduces the sample complexity prior to quantitation.

This method is ideal for complex samples, because only cysteine residues are tagged and labeled peptides are affinity purified, which significantly reduces sample complexity. ICAT labeling does have a bias against proteins and peptides that lack cysteine residues, which is considerable compared to proteins that lack lysine residues. For example, 14% of Escherichia coli (E. coli) open reading frames (ORFs) do not code for cysteines, while only 0.8% do not code for lysine (although half of those could still be tagged because of terminal amines). This difference in amino acid availability should be considered when determining the right isotopic labeling method to use for quantitative proteomic analyses. The group that originally developed ICAT reagents also later developed ICAT tags that contain ^{13}C instead of deuterium to circumvent the issue of partial peak separation during LC.

Although affinity purification of ICAT-labeled peptides reduces sample complexity by 10-fold, the cysteine-specific labeling method also reduces protein sequence coverage by the same factor. Because of this limitation, isotope-coded protein labeling (ICPL) was developed, in which lysine residues and available N-termini on intact proteins are isotopically labeled with a heavy (d4) or light (d0) tag. This approach increases the level of labeling, because significantly more terminal amino groups are available than cysteine resides. Also, ICPL is amenable to a greater level of pre-MS fractionation than other labeling methods, because sample complexity can be reduced at both the protein level (before digestion; electrophoresis or LC) and the peptide level (after digestion; LC). ICPL also allows the simultaneous comparison of three experimental conditions in a single experiment with two heavy tags (d7 and d3) and the d0 light tag. This multiplex capability distinguishes ICPL from ICAT and the other labeling methods listed above.

Isobaric Tags

Unlike isotopic tags that have the potential to separate during LC elution, isobaric tags have identical masses and chemical properties that allow heavy and light isotopologues to co-elute together. The tags are then cleaved from the peptides by collision-induced dissociation (CID) during MS/MS, which is required for this type of quantitative proteomic analysis. Indeed, these tags were originally called tandem mass tags to indicate their use with tandem mass spectrometry. After CID, the peptide fragment ions are analyzed for sequence assignment and the isobaric tags are quantitated, resulting in concurrent peptide identification and relative quantitation. Additionally, because MS/MS is required to detect the isobaric tags, unlabeled peptides are not quantitated.

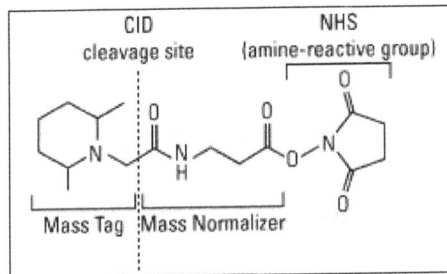

Structure of Tandem Mass Tags

Isobaric tags have ^{13}C and ^{15}N substitutions that give them variable masses. These differences are normalized by linkers that vary in mass depending on the mass of the tag. While this schematic shows N-hydroxysuccinimide, an amine-reactive moiety, sulfhydryl-reactive groups are also available to label cysteines.

A benefit of isobaric mass tags is the multiplex capabilities and thus increased throughput potential of this approach. Commercially available isobaric mass tags (e.g., TMT, iTRAQ) offer the simultaneously analysis of 4, 6 or 8 biological samples. While the exact tags used vary depending on manufacturer, the basic components of all isobaric mass tag reagents consist of a mass reporter (tag) that has a unique number of ^{13}C substitutions, a mass normalizer that has a unique mass that balances the mass of the tag to make all of the tags equal in mass. Isobaric mass tags also have a reactive moiety that crosslinks to primary amines or cysteines (depending on the product used). These tags are designed so that the mass tag is cleaved at a specific linker region upon high-energy CID (HCD), yielding the different sized tags that are then quantitated by LC-MS/MS. Isobaric mass tagging has also been adapted for use with protein labeling (similar to ICPL). Some commercially available kits also offer isobaric tags with sulfhydryl-reactivity and anti-TMT antibody for affinity purification of cysteine-tagged peptides prior to LC-MS/MS.

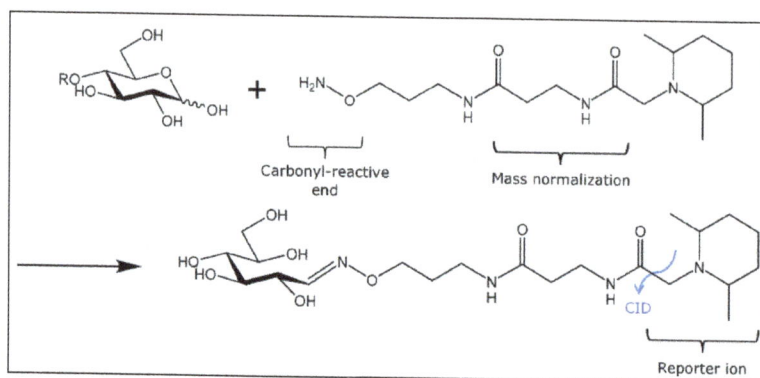

Example of Multiplex Proteomic Quantitation

Samples are labeled with individual mass tags and then combined for LC-MS/MS analysis. Because the masses of all of the tags are the same, identical peptides from different samples co-elute and are analyzed by MS. After HCD-induced tag cleavage and another round of MS, the tags are used to quantitate relative peptide intensities, while the peptide fragment ions are sequenced for protein identification.

Selected Reaction Monitoring and Targeted Assay Development

Selected reaction monitoring (SRM) or multiple reaction monitoring (MRM) is a method of absolute quantitation (AQUA) in targeted proteomics analyses that is performed by spiking complex samples with stable isotope-labeled synthetic peptides that act as internal standards for specific peptides. These heavy peptides are designed to be identical to tryptic peptides generated by sample digestion, so that they co-elute with the target peptide and are concomitantly analyzed by MS/MS (using instrumentation with a large dynamic range). The target peptide concentration is then determined by measuring the observed signal response for the target peptide relative to that of the heavy peptide, the concentration of which is calculated from a pre-determined calibration-response curve. While this method yields absolute peptide concentrations in as few as one sample, calibration curves have to be generated for each target peptide in the sample.

Assay development is a significant part of SRM proteomic analyses. Heavy peptides for each of the target peptides must be synthesized, and because proteins yield multiple peptides with varying electrochemical characteristics, the heavy peptide sequences that will yield the optimal results must be identified. Software is used to help predict the ideal tryptic peptide sequences, but the combination of trial-and-error peptide identification and instrumentation optimization makes absolute quantitation using isotopic peptides time consuming and costly. Once the assay is optimized for a predetermined set of peptides (up to approximately 200 per LC-MS run), though, SRM offers the highest level of reproducibility and sensitivity in detecting these peptides in multiple samples. This approach has been reported to detect proteins with concentrations less than 50 copies per cell in unfractionated lysates, demonstrating that it is the quantitative approach that is the least affected by sample complexity.

AQUA-grade peptides are costly because of their high quality and purity, and therefore scientists often use low-quality crude peptides during targeted assay development. Entire libraries of different peptide sequences can be commercially synthesized and screened during assay development to identify the optimum peptides, which are then synthesized at the AQUA purity and quality standards for SRM assays.

References

- Interaction-proteomics, proteomics: proteomics.case.edu, Retrieved 23 February, 2019

- Cohen, Philip (2002-05-01). "The origins of protein phosphorylation". Nature Cell Biology. 4 (5): E127–130. Doi:10.1038/ncb0502-e127. ISSN 1465-7392. PMID 11988757

- Quantitative-proteomics, pierce-protein-methods, protein-biology-resource-library, protein-biology-learning-center, life-science, home: thermofisher.com, Retrieved 24 March, 2019

- Berger AB, et al. Activity-based protein profiling: applications to biomarker discovery, in vivo imaging and drug discovery. American Journal of Pharmacogenomics 2004 Article

- Tribl F; K Marcus; G Bringmann; HE Meyer; M Gerlach; P Riederer (2006). "Proteomics of the Human Brain: Sub-Proteomes Might Hold the Key to Handle Brain Complexity". Journal of Neural Transmission. 113 (8): 1041–54. Doi:10.1007/s00702-006-0513-7. PMID 16835691

Protein Methods

CHAPTER 4

The techniques used to study proteins are known as protein methods. Some of these methods are protein purification, protein identification, protein mass spectrometry, protein sequencing, two-dimensional gel electrophoresis, threading, etc. The topics elaborated in this chapter will help in gaining a better perspective about these protein methods.

PROTEIN PURIFICATION

Protein purification is vital for the characterization of the function, structure, and interactions of proteins. The various steps in the purification process may include cell lysis, separating the soluble protein components from cell debris, and finally separating the protein of interest from product- and process-related impurities. Separation of the protein of interest, in its desired form, from all impurities, is typically the most challenging aspect of protein purification.

Affinity chromatography is an effective technique for protein purification that often enables a single-step purification of proteins to a purity level sufficient for analytical characterization. Affinity chromatography is a separation technique based on molecular conformation—molecules that "fit" one another bind selectively in a "lock and key" fashion (e.g., an antibody may recognize and specifically bind an antigen). The technique can use application-specific chromatography resins that have antibody ligands (fragments of antibodies that contain the antigen-binding domain) attached to the resin surface. Most frequently, these ligands function with the target protein in a manner similar to that of antibody-antigen interactions. This highly specific fit between the ligand and its target compound enables affinity column chromatography that is also highly specific. Antigens bind to the resin-bound antibody ligand, while other sample components and impurities do not bind and flow through the affinity column. Bound antigen (typically the protein of interest) can then be eluted, often with a pH change that breaks the molecular antigen–antibody interaction, yielding a single, highly pure elution peak.

Protein Purification Techniques

Gel Permeation Chromatography

Gel permeation chromatography separates protein molecules according to their size and shape, with large, nonspherical molecules eluting more rapidly than small, spherical molecules. Gel permeation resins may be thought of as beads containing numerous uniform conical pits on their surface. Molecules that are larger than the diameter of the pit do not enter it and consequently pass through the column quickly. Small molecules can enter the pits. The smaller the molecule, the deeper it can go into these conical pits and the longer it will remain there. Consequently. the presence of the pits impedes the progress of small molecules through the column more than it does the progress of large molecules.

In order to standardize a gel permeation column, it is necessary to calculate the fraction of the stationary gel volume that is available for diffusion of a given solute species (K_{av}) This is done by using the following parameters:

Elution volume (Ve)	-The volume at which a given protein elutes.
Total volume (Vt)	-Volume of the packed column bed.
Void volume (Vo)	-Elution volume of the molecules only distributed in the mobile phase of the gel because they are larger than the diameter of the largest pits in the gel.

$$K_{av} \frac{V_e - V_o}{V_t - V_o}$$

K_{av} for a given protein is proportional to the log of the molecular radius of that protein. Under optimal conditions, protein molecules do not bind to gel permeation resins. Both column length and flow rate affect the degree of resolution obtainable using gel permeation chromatography. Since resolution is directly proportional to the square root of the column length. a relatively longer column is desirable. Resolution is also improved by reducing the flow rate of the column. Because increasing column length and decreasing flow rate both increase the time required to elute the protein of interest from a gel permeation column, this increased time factor must be balanced against the increased resolution it provides.

Protein molecules bind electrostatically and reversibly to ion exchangers if their net charge is opposite to that of the exchanger. All proteins are amphoteric polyelectrolytes, which means that their net charge is dependent upon the pH of the solution in which they are suspended. At low pH, the net charge on most proteins is positive (binds to cation exchangers) and at high pH the net charge is usually negative (binds to anion exchangers). At the point of zero net charge, the isoelectric point, protein molecules are not bound to any type of ion exchanger. Since protein molecules can bind to either anion or to cation exchangers (depending upon pH), the stability of the protein at various pH's is usually the most important factor in choosing an ion exchanger for specific protein separations.

Ion-exchange Chromatography

An Ion exchanger Contains of an insoluble support matrix containing chemically bound positive (anion exchanger) or negative (anion exchanger) groups and mobile counter ions. The counter ions may be reversibly exchanged with other ions of the same charge without altering the insoluble matrix or the chemically bound groups.

Protein molecules bind electrostatically and reversibly to ion exchangers if their net charge is opposite to that of the exchanger. All proteins are amphoteric polyelectrolytes, which means that their net charge is dependent upon the pH of the solution in which they are suspended. At low pH, the net charge on most proteins is positive (binds to cation exchangers) and at high pH the net charge is usually negative (binds to anion exchangers). At the point of zero net charge, the isoelectric point, protein molecules are not bound to any type of ion exchanger. Since protein molecules can bind to either anion or to cation exchangers (depending upon pH), the stability of the protein at various pH's is usually the most important factor in choosing an ion exchanger for specific protein separations.

Once proteins have been bound to an ion exchange resin, selective elution can be accomplished in several ways. For example, altering the column buffer pH towards the isoelectric point of a specific protein will cause that protein to lose its net charge, desorb and elute from the column. Consequently, if a pH gradient is applied to an ion exchanger containing bound protein, each protein will elute as its isoelectric point is reached and protein separation will be accomplished. Changes in ionic strength of the column buffer can also be used to selectively release proteins from ion exchange resins. At low ionic strength, competition for charged groups on the ion exchanger is at a minimum and even slightly charged proteins are bound strongly. Increasing the ionic strength of the column buffer (by using a linear gradient) increases the competition and interferes with the interaction between the ion exchanger and the protein. Weakly charged molecules (those closest to their isoelectric pH) are eluted at lower ionic strength than more highly charged molecules.

Chromatofocusing is a specialized form of ion-exchange chromatography that requires a resin containing charged groups with high buffering capacity. The column is equilibrated at a given starting pH and protein is loaded. The column is then eluted with buffer of a different pH than the starting buffer. In this way, a pH gradient is formed , within the column as the eluting buffer titrates the ion exchanger. For example, in order to produce a descending pH gradient, (pH 9-6) the ion exchanger must be at a higher pH than the eluent. Consequently, a basic buffering group is required on the ion exchanger so an anion exchanger must be used. The column is first equilibrated at pH 9 and the pH of the eluent is set at pH 6. The eluent contains a large number of differently charged molecules and as they migrate down the column. the most acidic of these bind to the basic groups on the anion exchanger. In this way. the pH at successive points in the column is gradually lowered as elution proceeds and more eluent is added to the column. A given protein is released from a particular portion of the column when the pH in that region of the column reaches the isoelectric pH of the protein. The protein then moves through the column as additional exchanger is titrated to the isoelectric pH of the protein. Elution of different proteins is in descending order of their isoelectric points.

Hydrophobic Interaction Chromatography

In this type of chromatography, proteins are separated based on the strengths of their hydrophobic interactions with an uncharged resin containing hydrophobic groups. These interactions are facilitated by increasing ionic strength. Consequently, proteins are often adsorbed to a hydrophobic resin in the presence of high concentrations of neutral salt (e.g., NaCI).

Selective elution of adsorbed proteins is achieved by altering the eluent in a way that causes desorption based on the strength of the hydrophobic interaction of individual proteins with the hydrophobic matrix. This can be accomplished by lowering the ionic strength of the eluent, lowering the polarity of the eluent by including substances such as ethylene glycol, including detergent in the eluent, or raising the pH of the eluent.

Affinity Chromatography

Affinity chromatography is a type of adsorption chromatography in which the column bed material has biological affinity for the particular protein to be isolated. Specific adsorptive properties of the bed material are obtained by covalently coupling an appropriate binding ligand to an insoluble matrix. The covalently attached ligand then binds a specific protein from a complex mixture of proteins and the bound protein is retained on the matrix. Unbound proteins are then washed away and the bound protein is released by altering the composition of the eluent. Highly selective protein isolations are achieved due to natural biological specificities. For example. enzymes can be purified by using affinity columns containing a substrate analog, specific inhibitor or cofactor for that enzyme. Antigens can be purified by using affinity columns containing the appropriate antibody. Similarly, antibodies can be purified on affinity columns containing covalently bound antigen.

Affinity chromatography provides opportunities for the isolation of proteins based on their biological function and thus differs greatly from conventional chromatography techniques in which separation depends on gross physical and chemical differences between proteins. This is a powerful technique that is currently being utilized extensively in isolating and purifying proteins.

Types of Chromatographic Resins

 In addition to those already mentioned, many other resins have been successfully utilized in specific protein isolations. For example, both hydroxylapatite (15) and Cibacron Blue F3-GA (coupled to an insoluble matrix) (16) have been effective in specific protein purification protocols.

Preparative Isoelectric Focusing

Basically, this technique amounts to electrophoresis of proteins in a pH gradient. Small molecules with high buffering capacity are incorporated into a granulated gel (Sephadex

G75 or similar gel) and when a current is applied to the gel, these molecules migrate to their isoelectric pH (where they have zero net charge) and remain there. Because of the high buffering capacity of these small molecules, a pH gradient is formed in the gel. In this system, proteins move to their isoelectric pH and are thus separated according to their differing isoelectric points. The gel acts as a support matrix which holds the separated proteins in their respective positions once separation has occurred.

Once the individual proteins have been separated according to their isoelectric point, individual sections of the gel (about 1 cm wide) are removed. Buffer is added to these individual gel sections to make a slurry which is poured into a small column. Because most proteins are large enough to be completely excluded from the gel matrix, it is possible to remove the protein from the gel by washing this column with 2 to 3 volumes of the appropriate buffer.

This procedure provides a high-resolution method for purifying small quantities of a particular protein. One of the major limitations of this method is that many proteins precipitate at their isoelectric pH and this may interfere with separation.

High Performance Liquid Chromatography

Although this technique shows promise for the future it is not yet very practical for large scale preparation of large protein molecules It may be useful for specific applications, especially for small peptides. or as a substitute for preparative Isoelectric focusing It is very rapid and might be a possible alternative for purifying small quantities of an extremely labile protein.

Electrophoresis Procedures

Electrophoresis is a standard laboratory technique by which charged molecules are transported through a solvent by an electrical field. Both proteins and nucleic acids may be separated by electrophoresis, which is a simple, rapid, and sensitive analytical tool. Most biological molecules carry a net charge at any pH other than their isoelectric point and will migrate at a rate proportional to their charge density. The mobility of a molecule through an electric field will depend on the following factors: field strength, net charge on the molecule, size and shape of the molecule, ionic strength, and properties of the matrix through which the molecule migrates (e.g., viscosity, pore size). Polyacrylamide and agarose are two support matrices commonly used in electrophoresis. These matrices serve as porous media and behave like a molecular sieve. Agarose has a large pore size and is suitable for separating nucleic acids and protein complexes. Polyacrylamide has a smaller pore size and is ideal for separating most proteins and smaller nucleic acids.

Several forms of polyacrylamide gel electrophoresis (PAGE) exist, and each form can provide different types of information about proteins of interest. Denaturing and reducing sodium dodecyl sulfate (SDS)-PAGE with a discontinuous buffer system is the most widely used electrophoresis technique and separates proteins primarily by mass.

Nondenaturing PAGE, also called native PAGE, separates proteins according to their mass/charge ratio. Two-dimensional (2D) PAGE separates proteins by native isoelectric point in the first dimension and by mass in the second dimension.

SDS-PAGE separates proteins primarily by mass because the ionic detergent SDS denatures and binds to proteins to make them uniformly negatively charged. Thus, when a current is applied, all SDS-bound proteins in a sample will migrate through the gel toward the positively charged electrode. Proteins with less mass travel more quickly through the gel than those with greater mass because of the sieving effect of the gel matrix.

Once separated by electrophoresis, proteins can be detected in a gel with various stains, transferred onto a membrane for detection by western blotting and/or excised and extracted for analysis by mass spectrometry. Protein gel electrophoresis is, therefore, a fundamental step in many kinds of proteomics analysis.

Polyacrylamide Gels

Acrylamide is the material of choice for preparing electrophoretic gels to separate proteins by size. Acrylamide mixed with bisacrylamide forms a crosslinked polymer network when the polymerizing agent, ammonium persulfate (APS), is added. TEMED (N,N,N,N'-tetramethylenediamine) catalyzes the polymerization reaction by promoting the production of free radicals by APS.

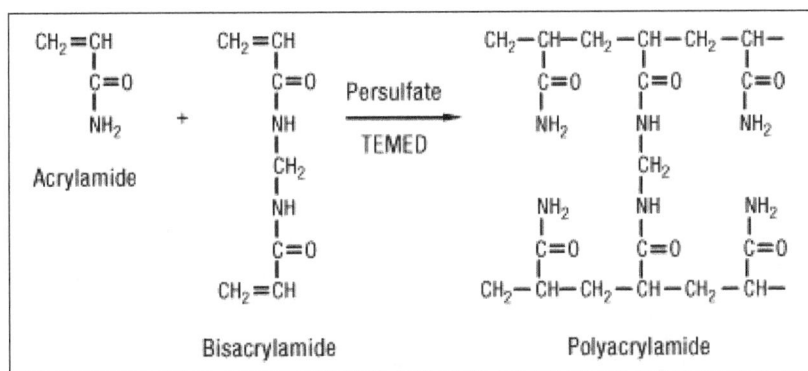

Polymerization and crosslinking of acrylamide. The ratio of bisacrylamide (N,N'-methylenediacrylamide) to acrylamide, as well as the total concentration of both components, affects the pore size and rigidity of the final gel matrix. These, in turn, affect the range of protein sizes (molecular weights) that can be resolved.

Example recipe for a traditional 10% Tris-glycine mini gel for SDS-PAGE:

- 7.5 mL 40% acrylamide solution

- 3.9 mL 1% bisacrylamide solution

- 7.5 mL 1.5 M Tris-HCl, pH 8.7

- Add water to 30 mL

- 0.3 mL 10% APS

- 0.3 mL 10% SDS

- 0.03 mL TEMED

The size of the pores created in the gel is inversely related to the polyacrylamide percentage (concentration). For instance, a 7% polyacrylamide gel has larger pores than a 12% polyacrylamide gel. Low-percentage gels are used to resolve large proteins, and high-percentage gels are used to resolve small proteins. "Gradient gels" are specially prepared to have a low percentage of acrylamide at the top (beginning of sample path) and a high percentage at the bottom (end), enabling a broader range of protein sizes to be separated.

Electrophoresis gels are formulated in buffers that enable electrical current to flow through the matrix. The prepared solution is poured into the thin space between two glass or plastic plates that form a "cassette". This process is referred to as "casting a gel". Once the gel polymerizes, the cassette is mounted (usually vertically) into an apparatus so that the top and bottom edges are placed in contact with buffer chambers containing a cathode and an anode, respectively. The running buffer contains ions that conduct current through the gel. When proteins are added in wells at the top edge and current is applied, the proteins are drawn by the current through the matrix slab and separated by the sieving properties of the gel.

To obtain optimal resolution of proteins, a "stacking" gel is cast over the top of the "resolving" gel. The stacking gel has a lower concentration of acrylamide (e.g., 7% for larger pore size), lower pH (e.g., 6.8), and a different ionic content. This allows the proteins in a loaded sample to be concentrated into one tight band during the first few minutes of electrophoresis before entering the resolving portion of a gel. A stacking gel is not necessary when using a gradient gel, as the gradient itself performs this function.

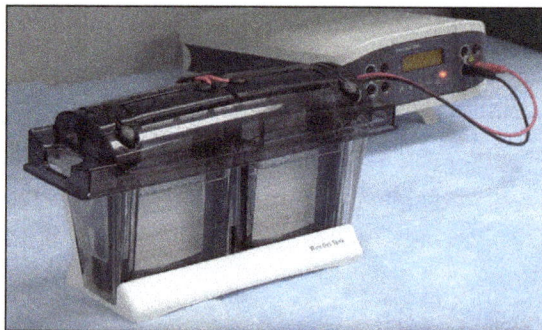

Polyacrylamide gel electrophoresis in progress.

Prepared gel cassettes are added to a gel tank, in this case the Invitrogen Mini Gel Tank, which holds two mini gels at a time. After wells are loaded with protein samples, the gels submerged in a conducting running buffer, and electrical current is applied, typically for 20 to 40 minutes. Run times vary according to the size and percentage of the gel and particular gel chemistry.

Precast Gels vs. Handcast Gels

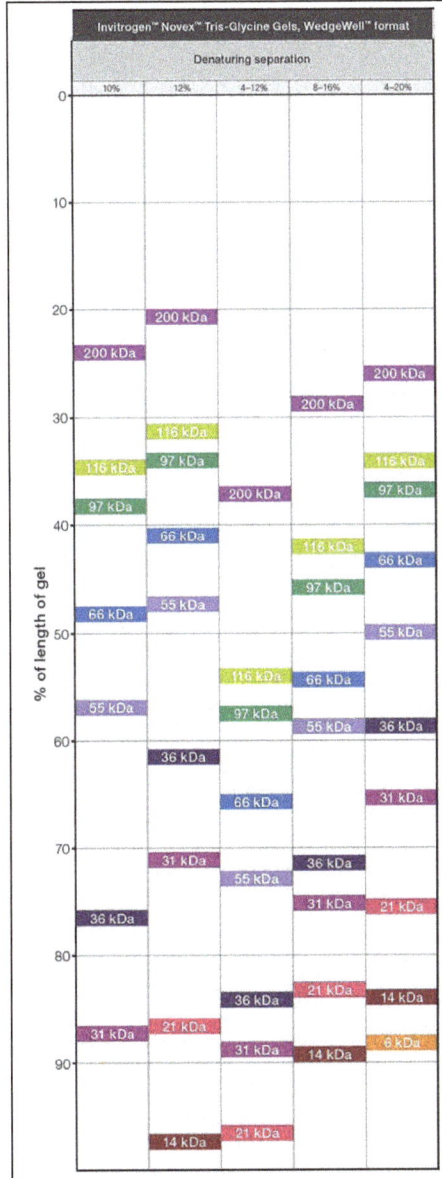

Example gel selection guide.

Traditionally, researchers casted their own gels using standard recipes that are widely available in protein methods literature. Most laboratories now depend on the convenience and consistency afforded by commercially available, ready-to-use precast gels. Precast gels are available in a variety of percentages, including difficult-to-pour gradient gels that provide excellent resolution and that separate proteins over the widest possible range of molecular weights. Precast gels are also available with several different buffer formulations (e.g., Tris-glycine, Tris-acetate, Bis-Tris), which are designed to optimize shelf life, run time, and protein resolution.

For researchers who require unique gel formulations not available as precast gels, a wide range of reagents and equipment are available. However, technological innovations in buffers and gel polymerization methods enable manufacturers to produce gels with greater uniformity and longer shelf life than individual researchers can prepare on their own with traditional equipment and methods. In addition, precast polyacrylamide gels eliminate the need to work with the acrylamide monomer, which is a known neurotoxin and suspected carcinogen.

Precast vs. handcast protein gels for SDS-PAGE. Polyacrylamide gels can be purchased precast and ready to use or prepared from reagents in the lab using a gel-casting system.

Many types of Invitrogen electrophoresis gels are available to enable researchers to separate proteins by different properties or for specific applications.

Denaturing vs. Native PAGE

SDS-PAGE and Denaturing Gels

SDS-PAGE is used for routine separation and analysis of proteins because of its speed, simplicity, and resolving capability. In SDS-PAGE, the gel is cast in a buffer containing sodium dodecyl sulfate (SDS), an anionic detergent. The protein samples are heated with SDS before electrophoresis so that the charge density of all proteins is made roughly equal. The SDS (aided by heat) denatures proteins in the sample and binds tightly to the uncoiled molecules. Usually, a reducing agent such as dithiothreitol (DTT) is also added to cleave protein disulfide bonds and ensure that no tertiary or quaternary protein structure remains. Consequently, when these samples are electrophoresed, proteins separate according to mass alone, with very little effect from compositional differences.

When a set of proteins of known mass are run alongside samples in the same gel, they provide a reference by which the mass of sample proteins can be determined. These sets of reference proteins are called mass markers or molecular weight markers (MW markers), protein ladders, or size standards, and they are available commercially in several forms.

Native PAGE

In native PAGE, proteins are separated according to the net charge, size, and shape of their native structure. Electrophoretic migration occurs because most proteins carry a net negative charge in alkaline running buffers. The higher the negative charge density (more charges per molecule mass), the faster a protein will migrate. At the same time, the frictional force of the gel matrix creates a sieving effect, regulating the movement of proteins according to their size and three-dimensional shape. Small proteins face only a small frictional force, while larger proteins face a larger frictional force. Thus native PAGE separates proteins based upon both their charge and mass.

Because no denaturants are used in native PAGE, subunit interactions within a multimeric protein are generally retained and information can be gained about the quaternary structure. In addition, some proteins retain their enzymatic activity (function) following separation by native PAGE. Thus, this technique may be used for preparation of purified, active proteins.

Following electrophoresis, proteins can be recovered from a native gel by passive diffusion or electro-elution. To maintain the integrity of proteins during electrophoresis, it is important to keep the apparatus cool and minimize denaturation and proteolysis. pH extremes should generally be avoided in native PAGE, as they may lead to irreversible damage, such as denaturation or aggregation, to proteins of interest.

1D vs. 2D PAGE

1-dimensional Polyacrylamide Gel Electrophoresis

The most common form of protein gel electrophoresis is comparative analysis of multiple samples by one-dimensional (1D) electrophoresis. Gel sizes range from 2 x 3 cm (tiny) to 15 x 18 cm (large format). The most popular size (approx. 8 x 8 cm) is usually referred to as a "minigel". Medium-sized gels (8 x 13 cm) are called "midi gels". Small gels require less time and reagents than their larger counterparts and are suited for rapid protein screening. However, larger gels provide better resolution and are needed for separating similar proteins or a large number of proteins

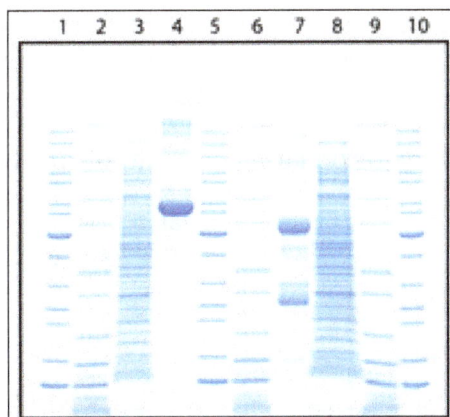

WedgWell Tris-Glycine gel.

Protein samples are added to sample wells at the top of the gel. When the electrical current is applied, the proteins move down through the gel matrix, creating what are called "lanes" of protein "bands". Samples that are loaded in adjacent wells and electrophoresed together are easily compared to each other after staining or other detection strategies. The intensity of staining and "thickness" of protein bands are indicative of their relative abundance. The positions (height) of bands within their respective lanes indicate their relative sizes (and/or other factors affecting their rate of migration through the gel).

Protein lanes and bands in 1D SDS-PAGE. Polyacrylamide gels may be formulated using a variety of different gel chemistries. Depicted here is a Protein ladders, purified proteins and E. coli lysate were loaded on a 4–20% gradient Novex WedgeWell Tris-Glycine gel; Lanes 1, 5, 10: 5 µL Thermo Scientific PageRuler Unstained Protein Ladder); lanes 2, 6, 9: 5 µL Mark12 Unstained Standard; lane 3: 10 µg E. coli lysate (10 µL sample volume); lane 4: 6 µg BSA (10 µL sample volume); lane 7: 6 µg hIgG (10 µL sample volume); lane 8: 20 µg E. coli lysate (20 µL sample volume). Electrophoresis was performed using the Mini Gel Tank. Sharp, straight bands were observed after staining with SimplyBlue SafeStain. Images were acquired using a flatbed scanner.

2-dimensional Polyacrylamide Gel Electrophoresis

Multiple components of a single sample can be resolved most completely by two-dimensional electrophoresis (2D-PAGE). The first dimension separates proteins according to their native isoelectric point (pI) using a form of electrophoresis called isoelectric focusing (IEF). The second dimension separates by mass using ordinary SDS-PAGE. 2D PAGE provides the highest resolution for protein analysis and is an important technique in proteomic research, where resolution of thousands of proteins on a single gel is sometimes necessary.

To perform IEF, a pH gradient is established in a tube or strip gel using a specially formulated buffer system or ampholyte mixture. Ready-made IEF strip gels (called immobilized pH gradient strips or IPG strips) and required instruments are available from certain manufacturers. During IEF, proteins migrate within the strip to become focused at the pH points at which their net charges are zero. These are their respective isoelectric points.

The IEF strip is then laid sideways across the top of an ordinary 1D gel, allowing the proteins to be separated in the second dimension according to size.

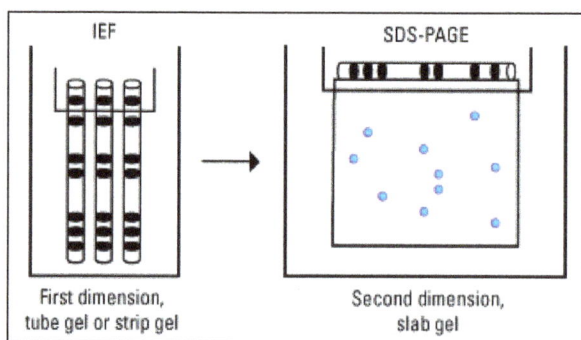

Overview of 2D gel electrophoresis. In the first dimension (left), one or more samples are resolved by isoelectric focusing (IEF) in separate tube or strip gels. IEF is usually performed using precast immobilized pH-gradient (IPG) strips on a specialized horizontal electrophoresis platform. For the second dimension (right), a gel containing the pI-resolved sample is laid across to top of a slab gel so that the sample can then be further resolved by SDS-PAGE.

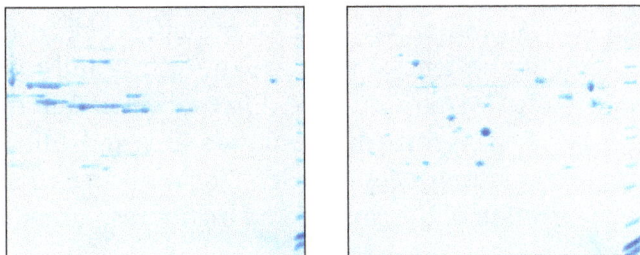

Analysis of spinach chloroplast extract by two-dimensional electrophoresis (2DE).

Spinach chloroplast extract was prefractionated in the ZOOM IEF Fractionator and the individual fractions were then separated by 2DE using narrow pH range ZOOM Strips and NuPAGE Novex 4–12% Bis-Tris ZOOM Gels. Gels were Coomassie stained using SimplyBlue SafeStain.

Protein Sample Preparation and Loading Buffers

Protein samples prepared for SDS-PAGE analysis are denatured by heating in the presence of a sample buffer containing 1% SDS with or without a reducing agent such as 20mM DTT, 2-mercaptoethanol (BME) or TCEP. The protein sample is mixed with the sample buffer and heated or boiled for 3 to 5 minutes (according to the specific protocol) then cooled to room temperature before it is pipetted into the sample well of a gel. Loading buffers also contain glycerol so that they are heavier than water and sink neatly to the bottom of the buffer-submerged well when added to a gel.

If a suitable, negatively charged, low-molecular weight dye is also included in the sample buffer, it will migrate at the buffer-front, enabling one to monitor the progress of electrophoresis. The most common tracking dye for sample loading buffers is bromophenol blue. Thermo Scientific Lane Marker Sample Buffers contain a bright pink tracking dye.

Samples may contain substances that interfere with electrophoresis by adversely affecting the migration of protein bands in the gel. Substances such as guanidine hydrochloride and ionic detergents can result in protein bands that appear smeared or wavy. The Thermo Scientific Pierce SDS-PAGE Sample Prep Kit facilitates removal of these interfering components using a specialized affinity resin system. Methods such as this are much faster than dialysis, ultrafiltration, or acetone precipitation, and the protein recovery is generally higher.

Running Buffers and Tanks

Protein gel electrophoresis power requirements as well as separation and migration patterns are determined by the chemical composition and pH of the buffer system.

Three basic types of buffers are required: the gel casting buffer, the sample buffer, and the running buffer that fills the electrode reservoirs. Electrophoresis may be performed using continuous or discontinuous buffer systems. A continuous buffer system, which utilizes only one buffer in the gel, sample, and gel chamber reservoirs, is most often used for nucleic acid analysis and rarely used for protein gel electrophoresis. Proteins separated using a continuous buffer system tend to be diffuse and poorly resolved. Conversely, discontinuous buffer systems utilize a different gel buffer and running buffer. These systems also use two gel layers of different pore sizes and different buffer compositions (the stacking and separating gels). Electrophoresis using a discontinuous buffer system results in concentration of the sample and higher resolution.

Comparison of Discontinuous Gel Systems

The classical Laemmli system, consisting of Tris-glycine gels and Tris-glycine running buffer, is employed for both SDS-PAGE and native PAGE and may be utilized to separate a broad range of proteins. This system is used widely because reagents for casting Tris-glycine gels are relatively inexpensive and readily available. Gels using this chemistry are also available in a variety of precast gel formats and percentages.

The formulation of this discontinuous buffer system creates a stacking effect to produce sharp protein bands at the beginning of the electrophoretic run. A boundary is formed between chloride, the leading ion, and glycinate, the trailing ion. Tris buffer provides the common cations. As proteins migrate into the resolving gel, they are separated according to size. Tris-glycine gels are used in conjunction with Laemmli sample buffer, and Tris/glycine/SDS running buffer is used for denaturing SDS-PAGE. Native PAGE is performed using native sample and running buffers without denaturants or SDS. The pH and ionic strength of the buffer used for running the gel (Tris, pH 8.3) are different from those of the buffers used in the stacking gel (Tris, pH 6.8) and the resolving gel (Tris, pH 8.8). The highly alkaline operating pH of the Laemmli system may cause band distortion, loss of resolution, or artifact bands.

- Hydrolysis of polyacrylamide at the high pH of the resolving gel, resulting in a short shelf life of 8 weeks.

- Chemical alterations such as deamination and alkylation of proteins due to the high pH of the resolving gel.

- Reoxidation of reduced disulfides from cysteine-containing proteins.

- Cleavage of Asp-Pro bonds of proteins when heated at 100°C in Laemmli sample buffer, pH 5.2.

Bis-tris

In contrast to conventional Tris-glycine gels, Bis-Tris HCl–buffered (pH 6.4) gels are more acidic, thus offering enhanced stability and greatly extended shelf-life over

Tris-glycine gels. For Bis-Tris gels, chloride serves as the leading ion and MES or MOPS act as the trailing ion. Bis-Tris buffer forms the common cation. Markedly different protein migration patterns are produced depending on whether a Bis-Tris gel is run with MES or MOPS denaturing running buffer: MES buffer is used for small proteins, and MOPS buffer is used for mid-sized proteins. Due to differences in ionic composition and pH, gel patterns obtained with Bis-Tris gels cannot be compared to those obtained with Tris-glycine gels. To prevent protein reoxidation, Bis-Tris gels must be run with alternative reducing agents such as sodium bisulfite. Reducing agents frequently used with Tris-glycine gels, such as beta-mercaptoethanol and dithiothreitol (DTT), do not undergo ionization at low pH levels and are not able to migrate with proteins in a Bis-Tris gel.

Tris-acetate

Tris-acetate gels are designed for separating large molecular weight proteins and may be used with both SDS-PAGE and native PAGE running buffers. Compared with Tris-glycine gels, Tris-acetate gels have a lower pH, which enhances the stability of these gels and minimizes protein modifications, resulting in sharper bands.

Tris Tricine

The Tris-Tricine gel system is a modification of the Tris-glycine gel system and is optimized to resolve low molecular weight proteins in the range of 2–20 kDa. As a result of reformulating the Laemmli running buffer and using Tricine in place of glycine, SDS-polypeptides form behind the leading ion front rather than running with the SDS front, thus allowing for their separation into discrete bands.

IEF

Isoelectric focusing (IEF) is a technique designed to separate proteins according to their isoelectric point (pI) rather than molecular weight. The pI is the pH at which a protein has no net charge and no longer moves in an electric field. IEF gels are run under native conditions and may be used for these applications: pI determination, detection of changes such as deamination, phosphorylation, or glycosylation, and resolution of different proteins similar in size that cannot be resolved using standard SDS-PAGE gels.

Zymogram

Zymogram gels are composed of gelatin or casein and are used to characterize proteases that utilize them as substrates. Samples are run under denaturing conditions, but due to the absence of reducing agents, proteins undergo renaturation. Proteolytic proteins present in the sample consume the substrate, generating clear bands against a background stained blue.

Protein Gel Electrophoresis Chambers

To perform protein gel electrophoresis, the polyacrylamide gel and buffer must be placed in an electrophoresis chamber that is connected to a power source, and which is designed to conduct current through the buffer solution. When current is applied, the smaller molecules migrate more rapidly and the larger molecules migrate more slowly through the gel matrix. Multiple gel chamber designs exist. The choice of equipment is usually based on these factors: the dimensions of the gel cassette, with some tank designs accommodating more cassette sizes than others; the nature of the protein target, and corresponding gel resolution requirements; and whether a precast or handcast gel, and vertical or horizontal electrophoresis system, has been selected.

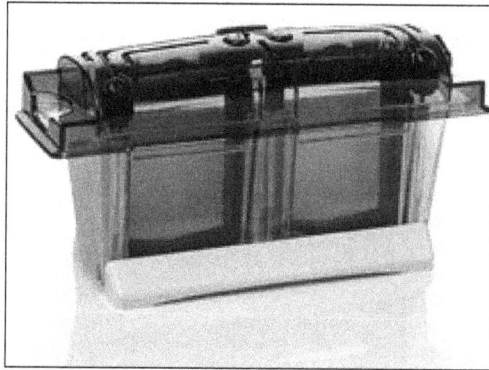

Mini gel tank for protein gel electrophoresis. This gel tank holds up to two mini gels and is compatible with the Invitrogen SureCast Gel Handcast System, and with NuPAGE, Bolt, or Novex mini gels. The unique tank design enables side-by-side gel loading and enhanced viewing during use.

Commercial gels are available in two size formats, for mini gels and midi gels. Both gels have similar run lengths, but midi gels are wider than mini gels, allowing midi gels to have more wells or larger wells. The additional wells in the midi gels permit more protein samples or large sample volumes to be loaded onto one gel.

Protein Ladders and Standards

To assess the molecular masses (sizes) of proteins in a gel, a prepared mixture containing several proteins of known molecular masses is run alongside the test sample in one or more lanes of the gel. Such sets of known proteins are called protein molecular weight (or mass) markers or protein ladders. A standard curve can be constructed from the distances migrated by each marker protein. The distance migrated by the unknown protein is then plotted, and the molecular weight is extrapolated from the standard curve.

Several kinds of ready-to-use protein molecular weight (MW) markers are available that are either unlabeled or prestained for different modes of detection. These are pre-reduced and, therefore, primarily suited for SDS-PAGE rather than native PAGE.

MW markers can also be made detectable via specialized labels, such as a fluorescent tag, and by other methods.

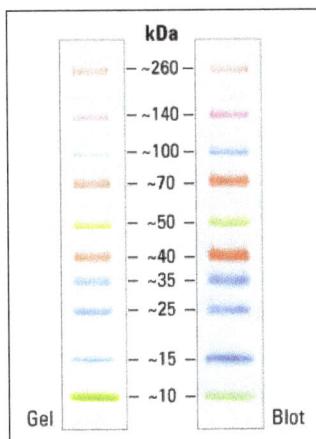

SDS-PAGE band profile of the Thermo Scientific PageRuler
Plus Prestained Protein Ladder. Images are from a 4–20%
Tris-glycine gel (SDS-PAGE) and subsequent transfer to a membrane.

Gel Staining

Once protein bands have been separated by polyacrylamide gel electrophoresis, they can be blotted (transferred) to a membrane for analysis by western blotting. Alternatively, they can be visualized directly in the gel using various staining or detection methods.

Coomassie dye is the most popular reagent for staining protein bands in electrophoretic gels. In acidic buffer conditions, Coomassie dye binds to basic and hydrophobic residues of proteins, changing in color from dull reddish-brown to intense blue. As with all staining methods, Coomassie dye reagents detect some proteins better than others based on their chemistry of action and differences in protein composition. For most proteins, however, Coomassie dye reagents detect as little as 10 ng per band in a mini gel.

In addition to the use of Coomassie dye, the traditional staining methods below may be employed for detection of protein bands in a gel:

Gel Staining Method	Considerations
Silver Staining	Metallic silver deposited onto the surface of the gel is used to detect protein bands. Method can detect ≤0.5 nanograms of protein and chemically crosslinks proteins in the gel matrix. Formulations exist that may be suitable for downstream applications, such as mass spectrometry.
Zinc reversible stains	Zinc does not stain the protein directly, but produces an opaque background with clear, unstained protein bands. Method detects ≤ 1 ng of protein and is reversible, allowing for analysis by mass spectrometry or western blotting.
Fluorescent Stains	Fluorescent stains are available with excitation and emission maxima corresponding to the common filter sets and laser settings of most fluorescence imagers.

Traditional staining methods involve one to several long incubation and wash steps. Recently, more rapid staining protocols have been developed using powered (electrophoretic) devices, such as the Thermo Scientific Pierce Power Stainer.

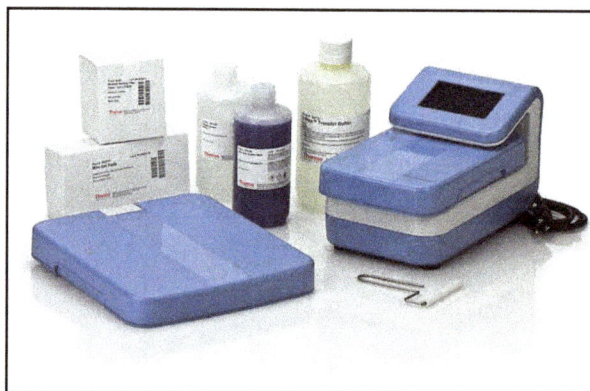

The Thermo Scientific Pierce Power Stainer. This powered device enables rapid (6–11 minute) Coomassie dye staining of proteins in polyacrylamide gels, including the removal of unbound stain, in a single step. The small, easy-to-use device consists of the Pierce Power Station and Pierce Power Stain Cassette, which accommodates up to two mini gels or one midi gel at a time. The staining procedure is designed exclusively for use with Pierce Power Staining Kits.

Transfer and Western Blots

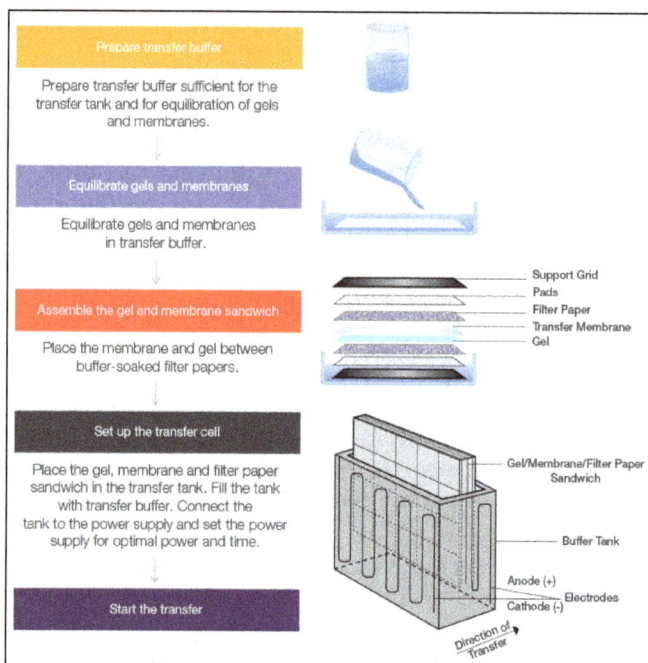

Following gel electrophoresis, proteins are transferred to a membrane in preparation for western blotting and other immunoblotting techniques.

Upon completion of protein gel electrophoresis, an immunodetection technique known as western blotting—also referred to as immunoblotting—may be performed. This technique involves the transfer of proteins that have been separated on a polyacrylamide gel to a solid support matrix, and subsequent detection of one or more specific proteins utilizing antigen-specific antibodies. Western blotting was introduced by Towbin et al. in 1979 and is now routinely used for protein analysis. Western blotting can produce qualitative and semiquantitative data about proteins of interest.

Scientists have used a variety of methods to transfer size-separated proteins from a polyacrylamide gel to the membrane support, including diffusion transfer, capillary transfer, heat-accelerated convectional transfer, vacuum blotting, and electroblotting (electrotransfer). Among these methods, electroblotting has emerged as the most popular because it is faster and more efficient than the other methods. The technique employs the use of various types of membranes, buffers, stains, molecular weight markers, and transfer devices. When setting up a western blot experiment, one of the first decisions is the choice of electrotransfer method: wet, semi-dry, or dry. This choice will drive what transfer buffers (or absence thereof) will be used and may also impact the choice of membrane products, because some transfer devices have specific requirements. The choice of transfer device may also drive the gel size, blotting area, and throughput, as well as potentially impact transfer efficiencies.

PROTEIN IDENTIFICATION

Protein occupies a special position in vivo, which is the main component of the protoplasm, so as nucleic acid, is the material basis of life phenomenon.As one of the material bases of life, proteins play a vital role in catalyzing the reactions of various organisms in vivo, regulating metabolism, resisting foreign invasion and controlling genetic information. Separation and characterization of proteins, quantitative analysis is the most important work in biochemistry and other biology, food testing, clinical testing, disease diagnosis, biopharmaceutical separation and purification and quality testing.

UV Absorption Method for Protein Estimation

This method estimates the amount of protein by measuring the characteristic absorption of tyrosine and tryptophan at 280 nm, which is simple, sensitive, fast, and still can be recycled after identification. In the protein molecule, the benzene ring of the tyrosine, phenylalanine and tryptophan residues contains conjugated double bonds, so that the protein has the property of absorbing UV. The absorbance will be proportional to the protein content while absorption peak is at 280nm. In addition, the protein solution at 238nm light absorption value and peptide bond content is proportional. The protein content can be measured by comparing the light absorption value of the protein solution with the protein concentration at a certain wavelength.

Coomassie Blue Staining for Protein Determination in Beer

Negatively-charged Coomassie brilliant blue dye binds to positively-charged proteins. When the dye is in solution, it's red and absorbs at 465 nm – but when it binds to basic amino acids in the protein, it becomes blue and absorbs at 595 nm. The absorption in your sample can then be compared to a standard curve. Coomassie blue includes G250 and R250, and G250 is commonly used for protein content identification as the binding reaction with protein is smooth. Although the reaction between R250 and G250 is relatively slow, it can be eluted so that can be used to stain the electrophoresis band.The reaction for this method is sensitive, which is 4 times higher than the Lowry method. The protein content of the microkine protein can be identified with this method, and the protein concentration is in the range of 0 ~ 1 000μg / mL.

High Pressure Liquid Chromatography

In recent years, high-pressure liquid chromatography technology has also been widely used in protein separation and determination. Due to the size of the protein, shape, charge, hydrophobicity, function and other characteristics of the protein, as well as the source and experimental requirements, there are different types of HPLC model can be utilized to separate the target protein. In order to detect more convenient and accurate methods for protein identification, scientist studied a variety of HPLC combined technology. Currently, the most commonly used combined technology includes HPLC-MS, HPLC-CE, HPLC-ITP, HPLC-ICP-AES.

More Protein Identification Methods

Besides these methods Haze-active Protein Determination Method, Kjeldahl determination, Enzyme Linked Immunosorbent Assay etc. These methods can be used to promote the accurate and improve efficiency for protein identification, being applied in various fields.

Top-down Proteomics

Top-down proteomics is a method of protein identification that either uses an ion trapping mass spectrometer to store an isolated protein ion for mass measurement and tandem mass spectrometry (MS/MS) analysis or other protein purification methods such as two-dimensional gel electrophoresis in conjunction with MS/MS. Top-down proteomics is capable of identifying and quantitating unique proteoforms through the analysis of intact proteins. The name is derived from the similar approach to DNA sequencing. During mass spectrometry intact proteins are typically ionized by electrospray ionization and trapped in a Fourier transform ion cyclotron resonance (Penning trap), quadrupole ion trap (Paul trap) or Orbitrap mass spectrometer. Fragmentation for tandem mass spectrometry is accomplished by electron-capture dissociation or electron-transfer dissociation. Effective fractionation is critical for sample handling before mass-spectrometry-based proteomics. Proteome analysis routinely involves digesting intact proteins followed by inferred protein identification using mass spectrometry (MS). Top-down MS (non-gel) proteomics interrogates protein structure through measurement of an intact mass followed by direct ion dissociation in the gas phase.

Advantages

- The main advantages of the top-down approach include the ability to detect degradation products, protein isoforms, sequence variants, combinations of post-translational modifications as well as simplified processes for data normalization and quantitation.

- Top-down proteomics, when accompanied with polyacrylamide gel electrophoresis, can help to complement the bottom-up proteomic approach. Top-down proteomic methods can assist in exposing large deviations from predictions and has been very successfully pursued by combining Gel Elution Liquid-based Fractionation Entrapment Electrophoresis fractionation, protein precipitation, and reverse phase HPLC with electrospray ionization and MS/MS.

- Characterization of small proteins represents a significant challenge for bottom up proteomics due to the inability to generate sufficient tryptic peptides for analysis. Top-down proteomics allows for low mass protein detection, thus increasing the repertoire of proteins known. While Bottom-up proteomics integrates cleaved products from all proteoforms produced by a gene into a single peptide map of the full-length gene product to tabulate and quantify expressed proteins, a major strength of Top-down proteomics is that it enables researchers to quantitatively track one or more proteoforms from multiple samples and to excise these proteoforms for chemical analysis.

Disadvantages

- In the recent past, the top down approach was relegated to analysis of individual proteins or simple mixtures, while complex mixtures and proteins were analyzed by more established methods such as Bottom-up proteomics. Additionally protein identification and proteoform characterization in the TDP (Top-down proteomics) approach can suffer from a dynamic range challenge where the same highly abundant species are repeatedly fragmented.

- Although Top-down proteomics can be operated in relatively high output in order to successfully map proteome coverage at a large level, the rate of identifying new proteins after initial rounds reduces quite sharply.

- Top-down proteomics interrogation can overcome problems for identifying individual proteins, but has not been achieved on a large scale due to a lack of intact protein fractionation methods that are integrated with tandem mass spectrometry.

PROTEIN MASS SPECTROMETRY

Mass spectrometry (MS) analysis of proteins measures the mass-to-charge ratio of ions to identify and quantify molecules in simple and complex mixtures. MS has become invaluable across a broad range of fields and applications, including proteomics. The development of high-throughput and quantitative MS proteomics workflows within the last two decades has expanded the scope of what we know about protein structure, function, modification and global protein dynamics.

Proteomics is the study of all proteins in a biological system (e.g., cells, tissue, organism) during specific biological events. Genomics and proteomics are considerably more difficult to study together than genomics or even transcriptomics alone, because of the dynamic nature of protein expression. Additionally, the majority of proteins undergo some form of posttranslational modification (PTM), further increasing proteomic complexity. During the last 15 years, the broad scope of proteomics is only beginning to be realized due in large part to technological developments in mass spectrometry.

Mass spectrometry is a sensitive technique used to detect, identify and quantitate molecules based on their mass-to-charge (m/z) ratio. Originally developed almost 100 years ago to measure elemental atomic weights and the natural abundance of specific isotopes, MS was first used in the biological sciences to trace heavy isotopes through biological systems. In later years, MS was used to sequence oligonucleotides and peptides and analyze nucleotide structure.

The development of macromolecule ionization methods, including electrospray ionization (ESI) and atmospheric pressure chemical ionization (APCI), enabled the study of protein structure by MS. Ionization also allowed scientists to obtain protein mass "fingerprints" that could be matched to proteins and peptides in databases and help identify unknown targets. New isotopic tagging methods led to the quantitation of target proteins both in relative and absolute quantities. All these technological advancements have resulted in methods that successfully analyze samples in solid, liquid or gas states. The sensitivity of current mass spectrometers allows one to detect analytes at concentrations in the attomolar range (10^{-18}).

Protein Mass Spectrometry Applications

Mass spectrometry measures the m/z ratio of ions to identify and quantify molecules in simple and complex mixtures. MS has become invaluable across a broad range of fields and applications, including proteomics. The development of high-throughput and quantitative MS proteomics workflows within the last two decades has expanded the scope of what we know about protein structure, function and modification, as well as global protein dynamics.

Common Applications and Fields that use Mass Spectrometry

Field of study	Applications
Proteomics	• Determine protein structure, function, folding and interactions.
	• Identify a protein from the mass of its peptide fragments.
	• Detect specific post-translational modifications throughout complex biological
	• mixtures.
	• Quantitate (relative or absolute) proteins in a given sample.
	• Monitor enzyme reactions, chemical modifications and protein digestion.
Drug Discovery	• Determine structures of drugs and metabolites.
	• Screen for metabolites in biological systems.
Clinical Testing	• Perform forensic analyses such as confirmation of drug abuse.
	• Detect disease biomarkers (e.g., newborns screened for metabolic diseases).
Genomics	• Sequence short oligonucleotides.
Environment	• Test food and beverages for contamination, adulteration and product consistency.
	• Analyze soil sample contaminants.
Geology	• Measure petroleum composition.
	• Perform carbon dating.

All mass spectrometers have an ion source, a mass analyzer and an ion detector. The nature of these components varies based on the purpose of the mass spectrometer, the type of data required, and the physical properties of the sample. Samples are loaded into the mass spectrometer in liquid, gas or dried form and then vaporized and ionized by the ion source (e.g., APCI, DART, ESI).

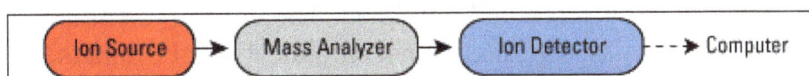

Schematic of the basic components of a mass spectrometer.

The charge that these molecules receive allows the mass spectrometer to accelerate the ions throughout the remainder of the system. The ions encounter electric and magnetic fields from mass analyzers, which deflect the paths of individual ions based on their m/z. Commonly used mass analyzers include time-of-flight [TOF], orbitraps, quadrupoles and ion traps, and each type has specific characteristics. Mass analyzers can be used to separate all analytes in a sample for global analysis, or they can be used like a filter to deflect only specific ions towards the detector.

Ions that have successfully been deflected by the mass analyzers then hit the ion detector. Most often, these detectors are electron multipliers or microchannel plates that emit a cascade of electrons when each ion hits the detector plate. This cascade results in amplification of each ion hit for improved sensitivity. This entire process is performed under an extreme vacuum (10^{-6} to 10^{-8} torr) to remove gas molecules and neutral and contaminating non-sample ions, which can collide with sample ions and alter their paths or produce non-specific reaction products.

Newer Thermo Scientific Orbitrap technology captures ions around a central spindle electrode and then analyzes their m/zvalues as they move across the spindle with different harmonic oscillation frequencies.

Mass spectrometers are connected to computer-based software platforms that measure ion oscillation frequencies and acquire mass spectra using image current detection. Data analysis programs detect ions and organize them by their individual m/z values and relative abundance. These ions can then be identified via established databases that predict the identity of the molecule based on its m/z value.

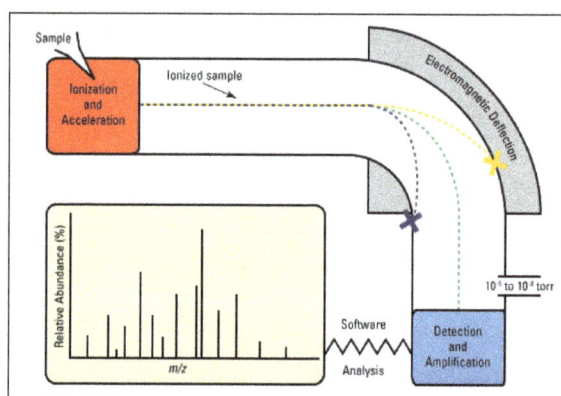

Diagram of a sector† mass spectrometer. A sample is injected into the mass spectrometer, and the molecules are ionized and accelerated. The ions are then separated by mass and charge by the mass analyzer via electromagnetic deflection, and the ions that are properly aligned are detected and amplified. The entire system is under intense vacuum during the entire process. After signal amplification, the data that is generated reports on the relative abundance of each ion based on its mass-to-charge (m/z) ratio. †Although sector instruments have decreased in use due to improvements in mass analyzers (e.g., quadrupole, orbitrap), this simplified diagram conveys a key principle of mass spectrometry, which is its ability to select and analyze specific ions in a complex sample.

Tandem Mass Spectrometry (MS/MS)

Tandem mass spectrometry (MS/MS) offers additional information about specific ions. In this approach, distinct ions of interest are in a quadrupole filter based on their m/z during the first round of MS and are fragmented by a number of different dissociation methods. One such method involves colliding the ions with a stream of inert gas, which is known as collision-induced dissociation (CID) or higher energy collision dissociation (HCD). Other methods of ion fragmentation include electron-transfer dissociation (ETD) and electron-capture dissociation (ECD).

These fragments are then separated based on their individual m/z ratios in a second round of MS. MS/MS is commonly used to sequence proteins and oligonucleotides because the fragments can be used to match predicted peptide or nucleic acid sequences, respectively, that are found in databases such as IPI, RefSeq and UniProtKB/Swiss-Prot. These sequence fragments can then be organized in silico into full-length sequence predictions.

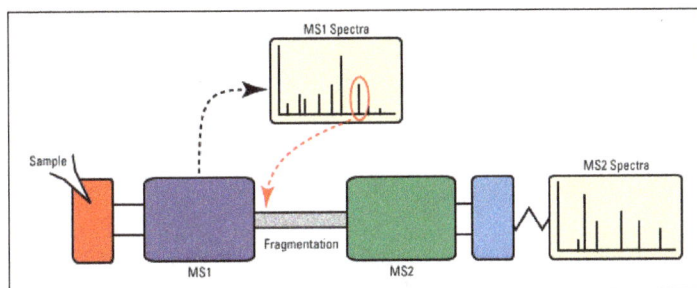

Diagram of tandem mass spectrometry (MS/MS). A sample is injected into the mass spectrometer, ionized, accelerated and analyzed by mass spectrometry (MS1). Ions from the MS1 spectra are then selectively fragmented and analyzed by a second stage of mass spectrometry (MS2) to generate the spectra for the ion fragments. While the diagram indicates separate mass analyzers (MS1 and MS2), some instruments utilize a single mass analyzer for both rounds of MS.

Biological samples are often quite complex and contain molecules that can mask the detection of the target molecule, such as when the sample exhibits a large dynamic concentration range between the target analytes and other molecules in the sample. Two methods of separation are commonly used to partition the target analytes from the other molecules in a sample.

Gas Chromatography and Liquid Chromatography Gas chromatography (GC) and liquid chromatography (LC) are common methods of pre-MS separation that are used when analyzing complex gas or liquid samples by MS, respectively. Liquid chromatography–mass spectrometry (LC-MS) is typically applied to the analysis of thermally unstable and nonvolatile molecules (e.g., sensitive biological fluids), while gas chromatography–mass spectrometry (GC-MS) is used for the analysis of volatile compounds such as petrochemicals. LC-MS and GC-MS also use different methods for ionization of the compound as it is introduced into the mass spectrometer. With LC-MS, electrospray ionization (ESI) is commonly applied, resulting in the production of aerosolized ions. With GC-MS, the sample may be ionized directly or indirectly via ESI.

High Performance Liquid ChromatographyHigh performance liquid chromatography (HPLC) is the most common separation method to study biological samples by MS or MS/MS (termed LC-MS or LC-MS/MS, respectively), because the majority of biological samples are liquid and nonvolatile. LC columns have small diameters (e.g., 75 μm; nanoHPLC) and low flow rates (e.g., 200 nL/min), which are ideal for minute samples. Additionally, "in-line" liquid chromatography (LC linked directly to MS) provides a high-throughput approach to sample analysis, enabling the elution of multiple analytes through the column at different rates to be immediately analyzed by MS. For example, 1 to 5 peptides in a complex biological mixture can be sequenced per second by in-line LC-MS/MS.

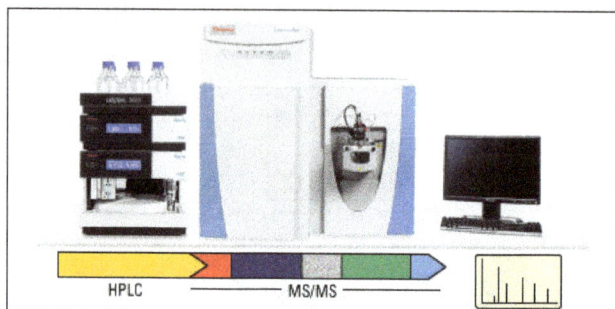

Example of in-line LC-MS/MS system. Thermo Scientific
Q Exactive Plus with Dionex UltiMate 3000 UHPLC.

PROTEIN SEQUENCING

Protein sequencing is the practical process of determining the amino acid sequence of all or part of a protein or peptide. This may serve to identify the protein or characterize its post-translational modifications. Typically, partial sequencing of a protein provides sufficient information (one or more sequence tags) to identify it with reference to databases of protein sequences derived from the conceptual translation of genes.

The two major direct methods of protein sequencing are mass spectrometry and Edman degradation using a protein sequenator (sequencer). Mass spectrometry methods are now the most widely used for protein sequencing and identification but Edman degradation remains a valuable tool for characterizing a protein's *N*-terminus.

Determining Amino Acid Composition

It is often desirable to know the unordered amino acid composition of a protein prior to attempting to find the ordered sequence, as this knowledge can be used to facilitate the discovery of errors in the sequencing process or to distinguish between ambiguous results. Knowledge of the frequency of certain amino acids may also be used to choose which protease to use for digestion of the protein. The misincorporation of low levels of non-standard amino acids (e.g. norleucine) into proteins may also be determined. A generalized method often referred to as *amino acid analysis* for determining amino acid frequency is as follows:

1. Hydrolyse a known quantity of protein into its constituent amino acids.

2. Separate and quantify the amino acids in some way.

Hydrolysis

Hydrolysis is done by heating a sample of the protein in 6 M hydrochloric acid to 100–110 °C for 24 hours or longer. Proteins with many bulky hydrophobic groups may require longer heating periods. However, these conditions are so vigorous that some amino acids (serine, threonine, tyrosine, tryptophan, glutamine, and cysteine) are degraded. To circumvent this problem, Biochemistry Online suggests heating separate samples for different times, analysing each resulting solution, and extrapolating back to zero hydrolysis time. Rastall suggests a variety of reagents to prevent or reduce degradation, such as thiol reagents or phenol to protect tryptophan and tyrosine from attack by chlorine, and pre-oxidising cysteine. He also suggests measuring the quantity of ammonia evolved to determine the extent of amide hydrolysis.

Separation and Quantitation

The amino acids can be separated by ion-exchange chromatography then derivatized to facilitate their detection. More commonly, the amino acids are derivatized then resolved by reversed phase HPLC.

An example of the ion-exchange chromatography is given by the NTRC using sulfonated polystyrene as a matrix, adding the amino acids in acid solution and passing a buffer of steadily increasing pH through the column. Amino acids are eluted when the pH reaches their respective isoelectric points. Once the amino acids have been separated, their respective quantities are determined by adding a reagent that will form a coloured derivative. If the amounts of amino acids are in excess of 10 nmol, ninhydrin can be used for this; it gives a yellow colour when reacted with proline, and a vivid purple with other amino acids. The concentration of amino acid is proportional to the absorbance of the resulting solution. With very small quantities, down to 10 pmol, fluorescent derivatives can be formed using reagents such as ortho-phthaldehyde (OPA) or fluorescamine.

Pre-column derivatization may use the Edman reagent to produce a derivative that is detected by UV light. Greater sensitivity is achieved using a reagent that generates a fluorescent derivative. The derivatized amino acids are subjected to reversed phase chromatography, typically using a C8 or C18 silica column and an optimised elution gradient. The eluting amino acids are detected using a UV or fluorescence detector and the peak areas compared with those for derivatised standards in order to quantify each amino acid in the sample.

N-terminal Amino Acid Analysis

Determining which amino acid forms the *N*-terminus of a peptide chain is useful for two reasons: to aid the ordering of individual peptide fragments' sequences into a whole chain, and because the first round of Edman degradation is often contaminated by impurities and therefore does not give an accurate determination of the *N*-terminal amino acid. A generalised method for *N*-terminal amino acid analysis follows:

1. React the peptide with a reagent that will selectively label the terminal amino acid.

2. Hydrolyse the protein.

3. Determine the amino acid by chromatography and comparison with standards.

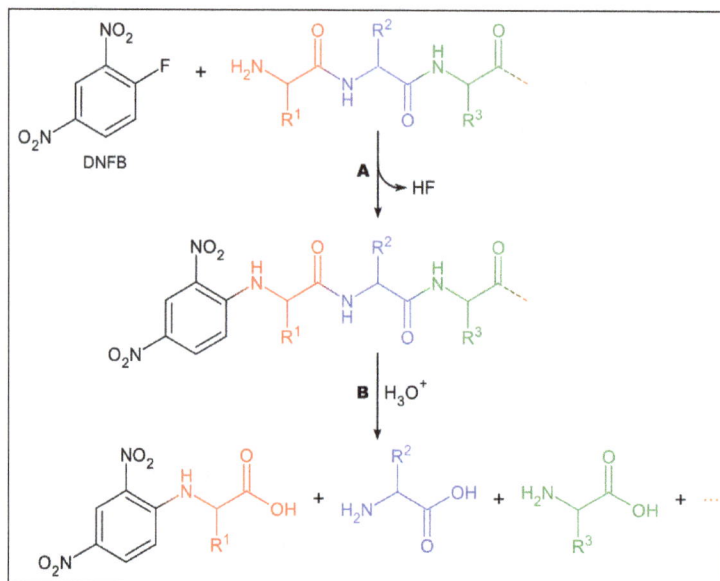

Sanger's method of peptide end-group analysis: A derivatization of *N*-terminal end with Sanger's reagent (DNFB), B total acid hydrolysis of the dinitrophenyl peptide.

There are many different reagents which can be used to label terminal amino acids. They all react with amine groups and will therefore also bind to amine groups in the side chains of amino acids such as lysine - for this reason it is necessary to be careful in interpreting chromatograms to ensure that the right spot is chosen. Two of the more common reagents are Sanger's reagent (1-fluoro-2,4-dinitrobenzene) and dansyl derivatives such as dansyl chloride. Phenylisothiocyanate, the reagent for the Edman degradation, can also be used. The same questions apply here as in the determination of amino acid composition, with the exception that no stain is needed, as the reagents produce coloured derivatives and only qualitative analysis is required. So the amino acid does not have to be eluted from the chromatography column, just compared with a standard. Another consideration to take into account is that, since any amine groups will have reacted with the labelling reagent, ion exchange chromatography cannot be used, and thin layer chromatography or high-pressure liquid chromatography should be used instead.

C-Terminal Amino Acid Analysis

The number of methods available for C-terminal amino acid analysis is much smaller than the number of available methods of N-terminal analysis. The most common method is to add carboxypeptidases to a solution of the protein, take samples at regular intervals, and determine the terminal amino acid by analysing a plot of amino acid concentrations against time. This method will be very useful in the case of polypeptides and protein-blocked N termini. C-terminal sequencing would greatly help in verifying the primary structures of proteins predicted from DNA sequences and to detect any postranslational processing of gene products from known codon sequences.

Edman Degradation

The Edman degradation is a very important reaction for protein sequencing, because it allows the ordered amino acid composition of a protein to be discovered. Automated Edman sequencers are now in widespread use, and are able to sequence peptides up to approximately 50 amino acids long. A reaction scheme for sequencing a protein by the Edman degradation follows; some of the steps are elaborated on subsequently.

1. Break any disulfide bridges in the protein with a reducing agent like 2-mercaptoethanol. A protecting group such as iodoacetic acid may be necessary to prevent the bonds from re-forming.

2. Separate and purify the individual chains of the protein complex, if there are more than one.

3. Determine the amino acid composition of each chain.

4. Determine the terminal amino acids of each chain.

5. Break each chain into fragments under 50 amino acids long.

6. Separate and purify the fragments.

7. Determine the sequence of each fragment.

8. Repeat with a different pattern of cleavage.

9. Construct the sequence of the overall protein.

Digestion into Peptide Fragments

Peptides longer than about 50-70 amino acids long cannot be sequenced reliably by the Edman degradation. Because of this, long protein chains need to be broken up into small fragments that can then be sequenced individually. Digestion is done either by endopeptidases such as trypsin or pepsin or by chemical reagents such as cyanogen bromide. Different enzymes give different cleavage patterns, and the overlap between fragments can be used to construct an overall sequence.

Reaction

The peptide to be sequenced is adsorbed onto a solid surface. One common substrate is glass fibre coated with polybrene, a cationic polymer. The Edman reagent, phenylisothiocyanate (PITC), is added to the adsorbed peptide, together with a mildly basic buffer solution of 12% trimethylamine. This reacts with the amine group of the N-terminal amino acid.

The terminal amino acid can then be selectively detached by the addition of anhydrous acid. The derivative then isomerises to give a substituted phenylthiohydantoin, which can be washed off and identified by chromatography, and the cycle can be repeated. The efficiency of each step is about 98%, which allows about 50 amino acids to be reliably determined.

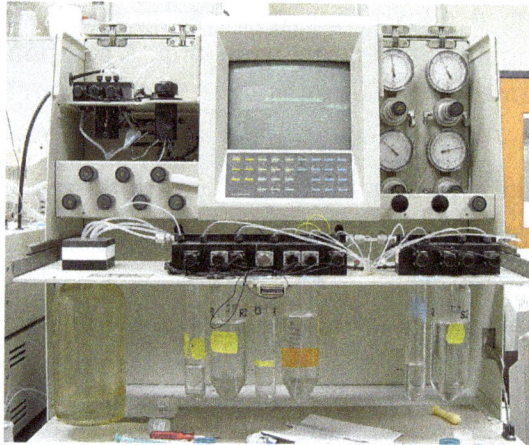

A Beckman-Coulter Porton LF3000G
protein sequencing machine

Protein Sequenator

A protein sequenator is a machine that performs Edman degradation in an automated manner. A sample of the protein or peptide is immobilized in the reaction vessel of the protein sequenator and the Edman degradation is performed. Each cycle releases and derivatises one amino acid from the protein or peptide's *N*-terminus and the released amino-acid derivative is then identified by HPLC. The sequencing process is done repetitively for the whole polypeptide until the entire measurable sequence is established or for a pre-determined number of cycles.

Identification by Mass Spectrometry

Protein identification is the process of assigning a name to a protein of interest (POI), based on its amino-acid sequence. Typically, only part of the protein's sequence needs to be determined experimentally in order to identify the protein with reference to databases of protein sequences deduced from the DNA sequences of their genes. Further protein characterization may include confirmation of the actual N- and C-termini of the POI, determination of sequence variants and identification of any post-translational modifications present.

Proteolytic Digests

A general scheme for protein identification is described.

- The POI is isolated, typically by SDS-PAGE or chromatography.

- The isolated POI may be chemically modified to stabilise Cysteine residues (e.g. S-amidomethylation or S-carboxymethylation).

- The POI is digested with a specific protease to generate peptides. Trypsin, which cleaves selectively on the C-terminal side of Lysine or Arginine residues, is the most commonly used protease. Its advantages include i) the frequency of Lys and Arg residues in proteins, ii) the high specificity of the enzyme, iii) the stability of the enzyme and iv) the suitability of tryptic peptides for mass spectrometry.

- The peptides may be desalted to remove ionizable contaminants and subjected to MALDI-TOF mass spectrometry. Direct measurement of the masses of the peptides may provide sufficient information to identify the protein but further fragmentation of the peptides inside the mass spectrometer is often used to gain information about the peptides' sequences. Alternatively, peptides may be desalted and separated by reversed phase HPLC and introduced into a mass spectrometer via an ESI source. LC-ESI-MS may provide more information than MALDI-MS for protein identification but uses more instrument time.

- Depending on the type of mass spectrometer, fragmentation of peptide ions may occur via a variety of mechanisms such as Collision-induced dissociation (CID) or Post-source decay (PSD). In each case, the pattern of fragment ions of a peptide provides information about its sequence.

- Information including the measured mass of the putative peptide ions and those of their fragment ions is then matched against calculated mass values from the conceptual (in-silico) proteolysis and fragmentation of databases of protein sequences. A successful match will be found if its score exceeds a threshold based on the analysis parameters. Even if the actual protein is not represented in the database, error-tolerant matching allows for the putative identification of a protein based on similarity to homologous proteins. A variety of software packages are available to perform this analysis.

- Software packages usually generate a report showing the identity (accession code) of each identified protein, its matching score, and provide a measure of the relative strength of the matching where multiple proteins are identified.

- A diagram of the matched peptides on the sequence of the identified protein is often used to show the sequence coverage (% of the protein detected as peptides). Where the POI is thought to be significantly smaller than the matched protein, the diagram may suggest whether the POI is an N- or C-terminal fragment of the identified protein.

De Novo Sequencing

The pattern of fragmentation of a peptide allows for direct determination of its sequence by *de novo* sequencing. This sequence may be used to match databases of protein sequences or to investigate post-translational or chemical modifications. It may provide additional evidence for protein identifications performed as above.

N- and C-termini

The peptides matched during protein identification do not necessarily include the N- or C-termini predicted for the matched protein. This may result from the N- or C-terminal peptides being difficult to identify by MS (e.g. being either too short or too long), being post-translationally modified (e.g. N-terminal acetylation) or genuinely differing from the prediction. Post-translational modifications or truncated termini may be identified by closer examination of the data (i.e. *de novo* sequencing). A repeat digest using a protease of different specificity may also be useful.

Post-translational Modifications

Whilst detailed comparison of the MS data with predictions based on the known protein sequence may be used to define post-translational modifications, targeted approaches to data acquisition may also be used. For instance, specific enrichment of phosphopeptides may assist in identifying phosphorylation sites in a protein. Alternative methods of peptide fragmentation in the mass spectrometer, such as ETD or ECD, may give complementary sequence information.

Whole-mass Determination

The protein's whole mass is the sum of the masses of its amino-acid residues plus the mass of a water molecule and adjusted for any post-translational modifications. Although proteins ionize less well than the peptides derived from them, a protein in solution may be able to be subjected to ESI-MS and its mass measured to an accuracy of 1 part in 20,000 or better. This is often sufficient to confirm the termini (thus that the protein's measured mass matches that predicted from its sequence) and infer the presence or absence of many post-translational modifications.

Limitations

Proteolysis does not always yield a set of readily analyzable peptides covering the entire sequence of POI. The fragmentation of peptides in the mass spectrometer often does not yield ions corresponding to cleavage at each peptide bond. Thus, the deduced sequence for each peptide is not necessarily complete. The standard methods of fragmentation do not distinguish between leucine and isoleucine residues since they are isomeric.

Because the Edman degradation proceeds from the N-terminus of the protein, it will not work if the N-terminus has been chemically modified (e.g. by acetylation or formation of Pyroglutamic acid). Edman degradation is generally not useful to determine the positions of disulfide bridges. It also requires peptide amounts of 1 picomole or above for discernible results, making it less sensitive than mass spectrometry.

Predicting from DNA/RNA Sequences

In biology, proteins are produced by translation of messenger RNA (mRNA) with the protein sequence deriving from the sequence of codons in the mRNA. The mRNA is itself formed by the transcription of genes and may be further modified. These processes are sufficiently understood to use computer algorithms to automate predictions of protein sequences from DNA sequences, such as from whole-genome DNA-sequencing projects, and have led to the generation of large databases of protein sequences such as UniProt. Predicted protein sequences are an important resource for protein identification by mass spectrometry.

Historically, short protein sequences (10 to 15 residues) determined by Edman degradation were back-translated into DNA sequences that could be used as probes or primers to isolate molecular clones of the corresponding gene or complementary DNA. The sequence of the cloned DNA was then determined and used to deduce the full amino-acid sequence of the protein.

Bioinformatics Tools

Bioinformatics tools exist to assist with interpretation of mass spectra , to compare or analyze protein sequences, or search databases using peptide or protein sequences.

PROTEIN ASSAYS METHODS

Protein concentration quantitation is an integral part of any laboratory workflow involving protein extraction, purification, labeling or analysis. Cell lysates are assayed to measure the protein yield from the lysis step and to normalize multiple samples for downstream application or for side-by-side comparison. Proteins obtained from a purification procedure are assayed to determine yield. Purified proteins that will be labeled with biotin or conjugated to reporter enzymes are typically assayed to ensure that the labeling reaction is prepared with appropriate stoichiometry. Given the wide range of reagent components that may be present in different kinds of samples, it is amazing that there exist protein assay reagents that are capable of reliably and specifically measuring the protein concentration.

Thermo Scientific Pierce Protein Assays provide a wide range of options for accurate protein concentration determination based on assay time, sensitivity, compatibility, standard curve linearity, and protein-to-protein variation.

Protein quantitation is often necessary before processing protein samples for isolation, separation and analysis by chromatographic, electrophoretic and immunochemical techniques. Depending on the accuracy required and the amount and purity of the protein available, different methods are appropriate for determining protein concentration.

The simplest and most direct assay method for proteins in solution is to measure the absorbance at 280 nm (UV range). Amino acids containing aromatic side chains (i.e., tyrosine, tryptophan and phenylalanine) exhibit strong UV-light absorption. Consequently, proteins and peptides absorb UV-light in proportion to their aromatic amino acid content and total concentration. Another method, traditionally used in amino acid analysis by HPLC, is to label all primary amines (i.e., N-terminus and side-chain of lysine residues) with a colored or fluorescent dye such as ninhydrin or o-phthalaldehyde (OPA). Direct UV-light absorbance and HPLC-reagent approaches have particular disadvantages that make them impractical for use with typical protein samples in proteomics workflows.

Instead, several colorimetric and fluorescent, reagent-based protein assay techniques have been developed that are used by nearly every laboratory involved in protein research. Protein is added to the reagent, producing a color change or increased fluorescence in proportion to the amount added. Protein concentration is determined by reference to a standard curve consisting of known concentrations of a purified reference protein. These protein assay techniques can be divided into two groups based on the type of chemistry involved.

Types and Examples of Protein Assay methods

Type: Assay based on	Example Thermo Scientific Pierce Protein Assays.
Protein-copper chelation and secondary detection of the reduced copper (Biuret method)	BCA and Rapid Gold BCA Modified Lowry.
Protein-dye binding and direct detection of the color change or increase in fluorescence associated with the bound dye	Coomassie (Bradford) Pierce 660 nm Quanti-iT Qubit EZQ.
Amine-reactive dye in the presence of cyanide or thiols, resulting in fluorescence.	CBQCAt.

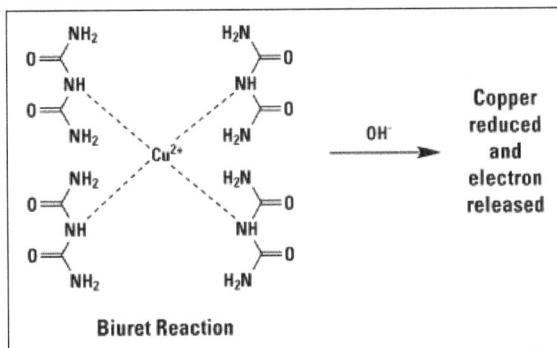

Biuret Reaction

Diagram of the biuret reaction.

By reducing the copper ion from cupric to cuprous form, the reaction produces a faint blue-violet color.

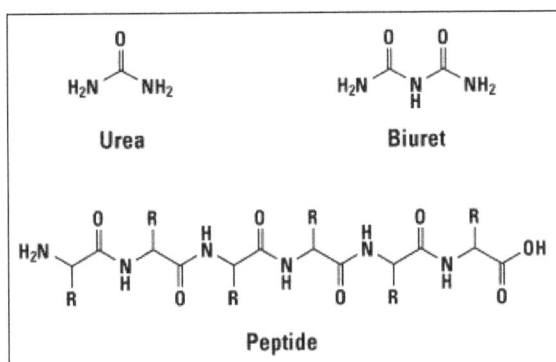

Structures of urea, biuret and peptide.

Because polypeptides have a structure similar to biuret, they are able to complex with copper by the biuret reaction.

Selecting a Protein Assay

No one reagent can be considered to be the ideal or best protein assay method. Each method has its advantages and disadvantages. The choice among available protein assays is usually based on the compatibility of the protein assay method with the samples. Additionally, one must consider potential interfering substances included in samples that may affect certain assay methods, as well as the accuracy, reproducibility and incubation time desired. Therefore, successful use of protein assays involves selecting the method that is most compatible with the samples to be analyzed, choosing an appropriate assay standard, and understanding and controlling the particular assumptions and limitations that remain.

The objective is to select a method that requires the least manipulation or pre-treatment of the samples to accommodate substances that interfere with the assay. Each method has its particular advantages and disadvantages. Because no one reagent can be considered the ideal or best protein assay method for all circumstances, most researchers have more than one type of protein assay available in their laboratories.

Important criteria for choosing an assay include:

- Compatibility with the sample type and components.

- Assay range and required sample volume.

- Protein-to-protein uniformity.

- Speed and convenience for the number of samples to be tested.

- Availability of spectrophotometer or plate reader necessary to measure the color produced (absorbance) by the assay.

The Pierce Rapid Gold BCA Protein Assay and Coomassie (Bradford) Protein Assay complement one another and provide the two basic methods for accommodating most samples. The various accessory reagents and alternative versions of these two assays accommodate many other particular sample needs.

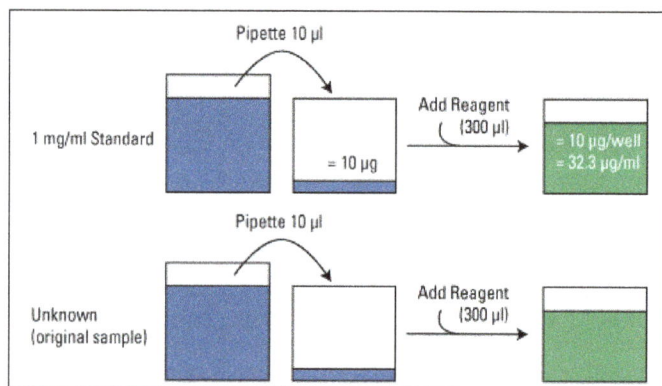

Diagram of protein assay steps.

If standard (top row) and unknown (bottom row) samples are dispensed and mixed with the same amount of assay reagent, then they are directly comparable. If the absorbances of the final solutions (green) are identical, then the concentration of the unknown sample is determined to be 1 mg/mL. The amount of protein in the assay well (middle) and the concentration in assay reagent (right) are irrelevant.

The Thermo Scientific Pierce Rapid Gold BCA Protein Assay was recently developed, and differs significantly from the conventional BCA protein assay. The colorimetric Pierce Rapid Gold assay retains the high sensitivity and linearity of the traditional BCA assay, but provides ready-to-read results within 5 minutes with room temperature (RT) incubation. In contrast, with traditional BCA assays—depending on the protocol—incubation times range from 30 minutes to 2 hours with temperatures ranging from 37° C to 60 °C.

Like the traditional BCA assay, the Pierce Rapid Gold BCA Protein assay involves the reduction of copper by proteins in an alkaline medium (biuret reaction) to produce sensitive and selective colorimetric detection by a new copper chelator. The amount of reduced copper is proportional to the amount of protein present in the solution. The selective copper chelator forms an orange-gold–colored complex that strongly absorbs light at 480 nm. This representative data compares the performance of the conventional and newly adapted BCA protein assays.

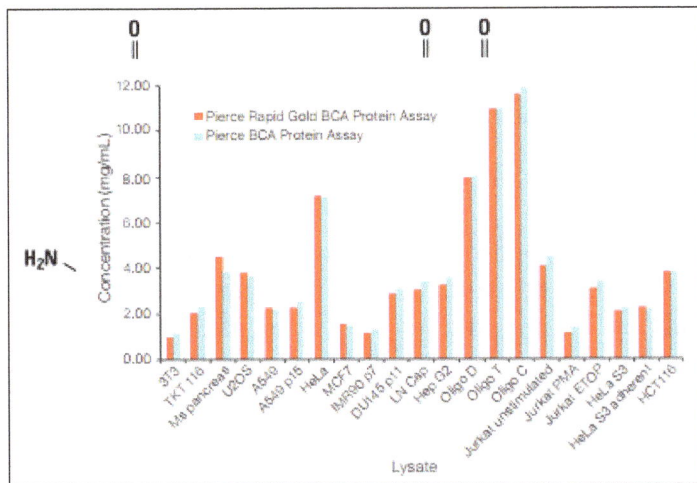

Protein concentration determination in lysates using the standard Pierce BCA Protein Assay and Pierce Rapid Gold BCA Protein Assay. Both assays were conducted according to the manufacturer's protocols, in a microplate format. For the standard BCA assay, 25 µL of sample was added to 200 µL of BCA working reagent and incubated for 30 minutes at 37 °C. For the Pierce Rapid Gold BCA Protein Assay, 20 µL of sample was added to 200 µL of the Pierce Rapid Gold BCA working reagent and incubated at room temperature for 5 minutes.

Selecting a Protein Standard

Because proteins differ in their amino acid compositions, each one responds somewhat differently in each type of protein assay. Therefore, the best choice for a reference standard is a purified, known concentration of the most abundant protein in the samples. This is usually not possible to achieve, and it is seldom convenient or necessary. In many cases, the goal is merely to estimate the total protein concentration, and slight protein-to-protein variability is acceptable.

If a highly purified version of the protein of interest is not available or it is too expensive to use as the standard, the alternative is to choose a protein that will produce a very similar color response curve in the selected protein assay method and is readily available to any laboratory at any time. Generally, bovine serum albumin (BSA) works well for a protein standard because it is widely available in high purity and relatively inexpensive. Alternatively, bovine gamma globulin (BGG) is a good standard when determining the concentration of antibodies because BGG produces a color response curve that is very similar to that of immunoglobulin G (IgG).

For greatest accuracy in estimating total protein concentration in unknown samples, it is essential to include a standard curve each time the assay is performed. This is particularly true for the protein assay methods that produce non-linear standard curves. Deciding on the number of standards and replicates used to define the standard curve depends upon the degree of non-linearity in the standard curve and the degree of accuracy required. In general, fewer points are needed to construct a standard curve if the color response is linear. Typically, standard curves are constructed using at least two replicates for each point on the curve.

Sample Preparation for Protein Assays

Before a sample is analyzed for total protein content, it must be solubilized, usually in a buffered aqueous solution. Additional precautions are often taken to inhibit microbial growth or to avoid casual contamination of the sample by foreign debris such as dust, hair, skin or body oils.

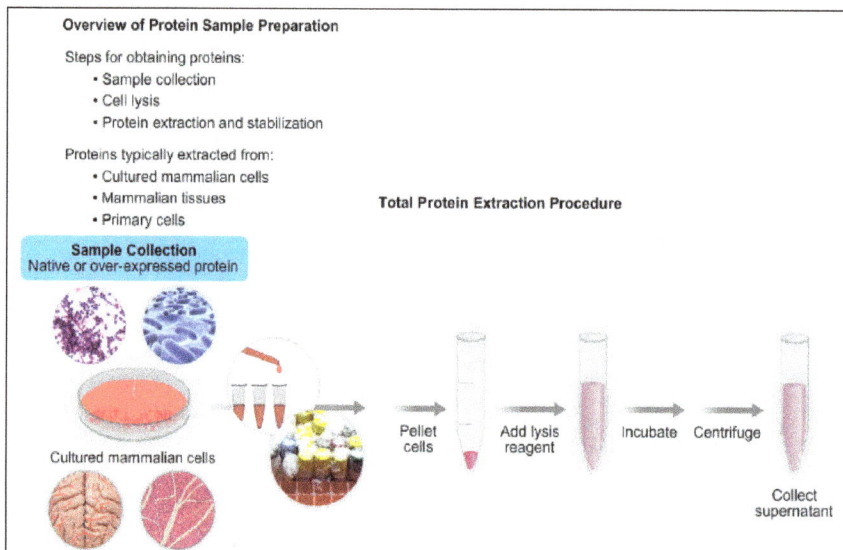

Detergent-based cell lysis. Both denaturing and non-denaturing cell lysis reagents may be used for protein extraction procedures.

Depending on the source material that the procedures involved before performing the protein assay, the sample will contain a variety of non-protein components. Awareness of these components is critical for choosing an appropriate assay method and evaluating the cause of anomalous results. For example, tissues and cells are usually lysed with buffers containing surfactants (detergents), biocides (antimicrobial agents) and protease inhibitors. Different salts, denaturants, reducing agents and chaotropes may also be included. After filtration or centrifugation to remove the cellular debris, typical samples will still include nucleic acids, lipids and other non-protein compounds.

Every type of protein assay is adversely affected by substances of one sort or another. Components of a protein solution are considered interfering substances in a protein assay if they artificially suppress the response, enhance the response, or cause elevated background by an arbitrarily chosen degree (e.g., 10% compared to control).

Inaccuracy resulting from a small amount of interfering substance can be eliminated by preparing the protein standard in the same buffer as the protein being assayed. For higher, incompatible levels of interfering substances, other strategies are necessary:

- Choose a different protein assay method or a version of the same assay method that includes components to overcome the interference.

- Dialyze or desalt the sample to remove interfering substances that are small (i.e., less than 1000 daltons), such as reducing agents.

- Precipitate the protein in TCA or other appropriate reagent, remove the solution containing the interfering component, and then re-dissolve the protein for analysis. This illustration provides an overview of how protein dialysis methods are used to remove substances that may contaminate protein samples and interfere with downstream applications.

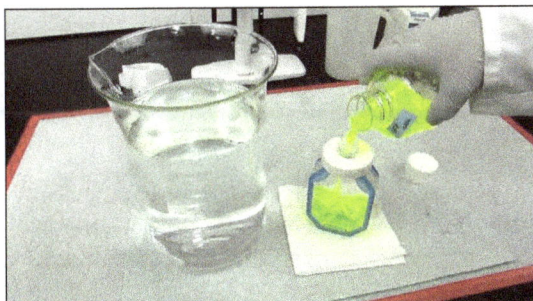

The schematic here shows how a dialysis cassette can be used for protein cleanup. 3 mL of 1 mg/mL IgG in 0.1 M Tris buffer, pH 7 inside a dialysis cassette is placed in 1,000 mL of 100 mM PBS, with a pH of 7.6. The old dialysate is discarded and replaced with 1,000 mL of 100 mM PBS, with a pH of 7.6. IgG is too large to enter the pores in the membrane; therefore, the amount of IgG inside the cassette remains constant. The Tris buffer concentration drops below 0.01 M inside the cassette as the Tris buffer diffuses out and the PBS buffer diffuses in. Again, the old dialysate is discarded and replaced with 1,000 mL of 100 mM PBS, with a pH of 7.6. The IgG inside of the cassette remains constant. The Tris buffer inside of the cassette drops to near undetectable levels. The buffer inside the cassette is 100 mM PBS, with a pH of 7.6.

High protein recovery is obtained using the 2 mL Thermo Scientific Slide-A-Lyzer MINI Dialysis Device.

Membrane MWCO (K)	Protein/Peptide	Recovery (%)
3.5	Insulin Chain B (3.5 kDa)	90.13
10	Cytochrome C (12.4 kDa)	94.44
20	Myoglobin (17 kDa)	95

Insulin chain B, cytochrome C and myoglobin (0.25 mg/mL) in either 50mM sodium phosphate, 75 mM NaCl at pH 7.2 or 0.2 M carbonate-bicarbonate buffer at pH 9.4 were dialyzed overnight (17 hours) at 4°C. The amount of protein in the retentate was determined using the Pierce BCA Protein Assay (Product # 23225).

Protein-to-protein Variation

Each protein in a sample responds uniquely in a given protein assay. Such protein-to-protein variation refers to differences in the amount of color (absorbance) obtained when the same mass of various proteins is assayed concurrently by the same method. These differences in color response relate to differences in amino acid sequence, isoelectric point (pI), secondary structure and the presence of certain side chains or prosthetic groups.

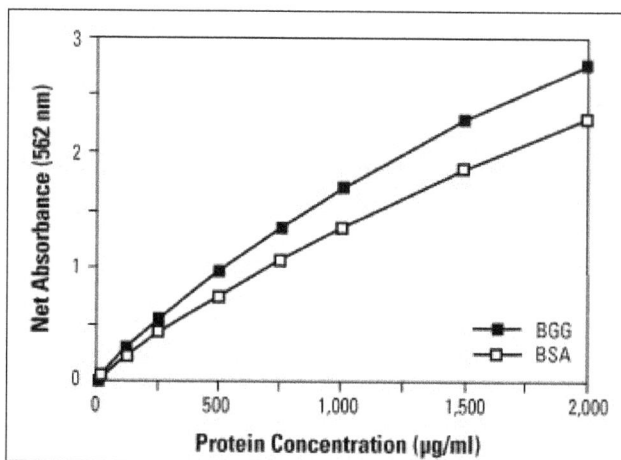

Standard curves: Typical standard curves for bovine serum albumin (BSA) and bovine gamma globulin (BGG) in the BCA Protein Assay. Kits include ampules of Albumin Standard.

Example standard curves for the Thermo Scientific Pierce BCA Protein Assay Kit. Eight concentrations of bovine serum albumin (BSA) and bovine gamma globulin (BGG) were assayed. The response values (absorbances) were plotted and a best-fit line drawn through the points. If unknown samples had been tested at the same time, their concentrations could be determined by reference to the one of these standard curves.

Depending on the sample type and purpose for performing an assay, protein-to-protein variation is an important consideration in selecting a protein assay method and in selecting an appropriate assay standard (e.g., BSA vs. BGG). Protein assay methods based on similar chemistry have similar protein-to-protein variation.

Protein-protein Variation of Protein Assays

Protein assays differ in their chemical basis for detecting protein-specific functional groups. Some assay methods detect peptide bonds, but no assay does this exclusively. Instead, each protein assay detects one or several different particular amino acids with greater sensitivity than others. Consequently, proteins with different amino acid compositions produce color at different rates or intensities in any given protein assay.

The following table compares the protein-to-protein variability in color respons general guideline for evaluating response differences among protein samples. However, because the comparisons were made using one protein concentration and buffer, they should not be used as exact calibration factors.

This variability information is helpful for choosing a protein standard. For example, when the sample to be assayed is a purified antibody, bovine gamma globulin (BGG, protein #5) will be a more accurate standard than bovine serum albumin (BSA, protein #1). These data also indicate the importance of specifying which assay standard was used when reporting protein assay results.

Calculations and Data Analysis

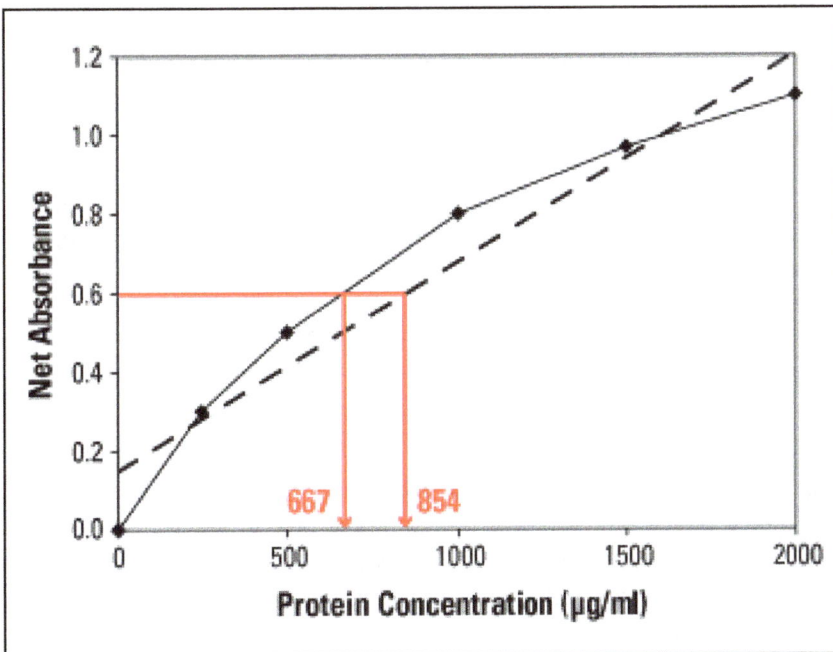

Comparison of point-to-point and linear-fit standard curves.

Interpolation and calculation for a test sample having absorbance 0.6 results in significantly different protein concentration values. In this case, the point-to-point method clearly provides a more accurate reference line for calculating the test sample.

With most protein assays, sample protein concentrations are determined by comparing their assay responses to that of a dilution-series of standards whose concentrations are known. Protein samples and standards are processed in the same manner by mixing them with assay reagent and using a spectrophotometer to measure the absorbances. The responses of the standards are used to plot or calculate a standard curve. Absorbance values of unknown samples are then interpolated onto the plot or formula for the standard curve to determine their concentrations.

Obviously, the most accurate results are possible only when unknown and standard samples are treated identically. This includes assaying them at the same time and in the same buffer conditions, if possible. Because different pipetting steps are involved, replicates are necessary if one wishes to calculate statistics (e.g., standard deviation, coefficient of variation) to account for random error.

Although most modern spectrophotometers and plate readers have built-in software programs for protein assay data analysis, several factors are frequently misunderstood by technicians. Taking a few minutes to study and correctly apply the principals involved in these calculations can greatly enhance one's ability to design assays that yield the most accurate results possible.

Peptide Quantitation Assays

For workflows utilizing proteomics using mass spectrometry, it is important to measure peptide concentration following protein digestion, enrichment, and C18 clean-up steps in order to normalize.

Sample-to-sample variation. In particular, for experiments utilizing isobaric labeling, it is critical to ensure that equal amounts of sample are labeled before mixing in order to have accurate results.

Similar to protein assay methods, various options are available for determining peptide concentration. Historically, UV-Vis (A280) or colorimetric, reagent-based protein assay techniques have been employed to measure peptide concentrations. Both BCA and micro-BCA assays are frequently used. Although these strategies work well for protein samples, these reagents are not designed for accurately detecting peptides. Alternatively, quantitative peptide assays— in either a colorimetric or flurometric format—are available to specifically quantitate peptide mixtures. When deciding to use a colormetric or fluorometric microplate assay format for quantitative peptide assays these important criteria must be considered:

- Compatibility with the sample type, components, and workflows.

- Assay range and required sample volume.

- Speed and convenience for the number of samples to be tested.

- Availability of the spectrophotometer or fluorometer needed to measure the output of the assay.

This representative data compares results obtain using colorimetric and fluorometric assays.

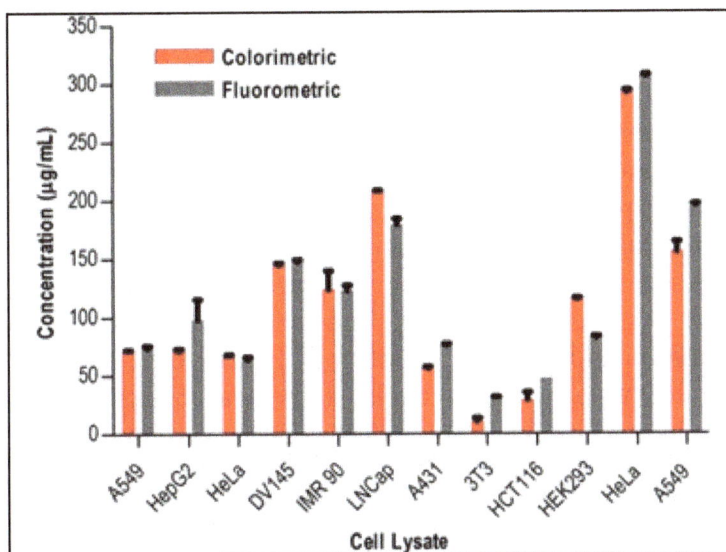

Quantitation comparison between colorimetric and fluorometric peptide assays. Tryptic peptide digests were prepared from twelve cell lines. Peptide digest concentrations were determined using the Thermo Scientific Pierce Quantitative Colorimetric Peptide Assay and the Pierce Quantitative Fluorometric Peptide Assay Kits according to instructions. Each sample was assayed in triplicate, and the concentration of each digest was calculated with standard curve generated using the Protein Digest Assay Standard.

TWO-DIMENSIONAL GEL ELECTROPHORESIS

Two-dimensional gel electrophoresis, abbreviated as 2-DE or 2-D electrophoresis, is a form of gel electrophoresis commonly used to analyze proteins. Mixtures of proteins are separated by two properties in two dimensions on 2D gels. 2-DE was first independently introduced by O'Farrell and Klose in 1975.

Basis for Separation

2-D electrophoresis begins with electrophoresis in the first dimension and then separates the molecules perpendicularly from the first to create an electropherogram in the second dimension. In electrophoresis in the first dimension, molecules are separated linearly according to their isoelectric point. In the second dimension, the molecules are then separated at 90 degrees from the first electropherogram according to molecular mass. Since it is unlikely that two molecules will be similar in two distinct properties, molecules are more effectively separated in 2-D electrophoresis than in 1-D electrophoresis.

The two dimensions that proteins are separated into using this technique can be isoelectric point, protein complex mass in the native state, or protein mass.

Separation of the proteins by isoelectric point is called isoelectric focusing (IEF). Thereby, a gradient of pH is applied to a gel and an electric potential is applied across the gel, making one end more positive than the other. At all pH values other than their isoelectric point, proteins will be charged. If they are positively charged, they will be pulled towards the more negative end of the gel and if they are negatively charged they will be pulled to the more positive end of the gel. The proteins applied in the first dimension will move along the gel and will accumulate at their isoelectric point; that is, the point at which the overall charge on the protein is 0 (a neutral charge).

For the analysis of the functioning of proteins in a cell, the knowledge of their cooperation is essential. Most often proteins act together in complexes to be fully functional. The analysis of this sub organelle organisation of the cell requires techniques conserving the native state of the protein complexes. In native polyacrylamide gel electrophoresis (native PAGE), proteins remain in their native state and are separated in the electric field following their mass and the mass of their complexes respectively. To obtain a separation by size and not by net charge, as in IEF, an additional charge is transferred to the proteins by the use of Coomassie Brilliant Blue or lithium dodecyl sulfate. After completion of the first dimension the complexes are destroyed by applying the denaturing SDS-PAGE in the second dimension, where the proteins of which the complexes are composed of are separated by their mass.

Before separating the proteins by mass, they are treated with sodium dodecyl sulfate (SDS) along with other reagents (SDS-PAGE in 1-D). This denatures the proteins (that is, it unfolds them into long, straight molecules) and binds a number of SDS molecules roughly proportional to the protein's length. Because a protein's length (when unfolded) is roughly proportional to its mass, this is equivalent to saying that it attaches a number of SDS molecules roughly proportional to the protein's mass. Since the SDS molecules are negatively charged, the result of this is that all of the proteins will have approximately the same mass-to-charge ratio as each other. In addition, proteins will not migrate when they have no charge (a result of the isoelectric focusing step) therefore the coating of the protein in SDS (negatively charged) allows migration of the proteins in the second dimension (SDS-PAGE, it is not compatible for use in the first dimension as it is charged and a nonionic or zwitterionic detergent needs to be used). In the second dimension, an electric potential is again applied, but at a 90 degree angle from the first field. The proteins will be attracted to the more positive side of the gel (because SDS is negatively charged) proportionally to their mass-to-charge ratio. As previously explained, this ratio will be nearly the same for all proteins. The proteins' progress will be slowed by frictional forces. The gel therefore acts like a molecular sieve when the current is applied, separating the proteins on the basis of their molecular weight with larger proteins being retained higher in the gel and smaller proteins being able to pass through the sieve and reach lower regions of the gel.

Detecting Proteins

The result of this is a gel with proteins spread out on its surface. These proteins can then be detected by a variety of means, but the most commonly used stains are silver and Coomassie Brilliant Blue staining. In the former case, a silver colloid is applied to the gel. The silver binds to cysteine groups within the protein. The silver is darkened by exposure to ultra-violet light. The amount of silver can be related to the darkness, and therefore the amount of protein at a given location on the gel. This measurement can only give approximate amounts, but is adequate for most purposes. Silver staining is 100x more sensitive than Coomassie Brilliant Blue with a 40-fold range of linearity.

Molecules other than proteins can be separated by 2D electrophoresis. In supercoiling assays, coiled DNA is separated in the first dimension and denatured by a DNA intercalator (such as ethidium bromide or the less carcinogenic chloroquine) in the second. This is comparable to the combination of native PAGE /SDS-PAGE in protein separation.

Common Techniques

IPG-DALT

A common technique is to use an Immobilized pH gradient (IPG) in the first dimension. This technique is referred to as IPG-DALT. The sample is first separated onto IPG gel (which is commercially available) then the gel is cut into slices for each sample which is then equilibrated in SDS-mercaptoethanol and applied to an SDS-PAGE gel for resolution in the second dimension. Typically IPG-DALT is not used for quantification of proteins due to the loss of low molecular weight components during the transfer to the SDS-PAGE gel.

2D Gel Analysis Software

In quantitative proteomics, these tools primarily analyze bio-markers by quantifying individual proteins, and showing the separation between one or more protein "spots" on a scanned image of a 2-DE gel. Additionally, these tools match spots between gels of similar samples to show, for example, proteomic differences between early and advanced stages of an illness. Software packages include Delta2D, ImageMaster, Melanie, PDQuest, Progenesis and REDFIN – among others. While this technology is widely utilized, the intelligence has not been perfected. For example, while PDQuest and Progenesis tend to agree on the quantification and analysis of well-defined well-separated protein spots, they deliver different results and analysis tendencies with less-defined less-separated spots.

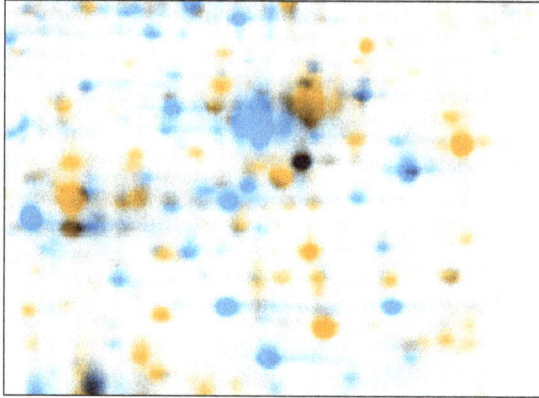

Warping: Images of two 2D electrophoresis
gels, overlaid with Delta2D.

First image is colored in orange, second one colored in blue. Due to running differences, corresponding spots do not overlap.

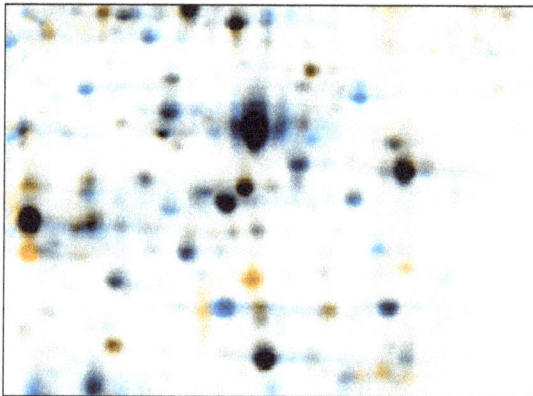

Warping: Images of two 2D electrophoresis
gels after warping.

First image is colored in orange, second one colored in blue. Corresponding spots overlap after warping. Common spots are colored black, orange spots are only present (or much stronger) on the first image, blue spots are only present (or much stronger) on the second image.

Challenges for automatic software-based analysis include incompletely separated (overlapping) spots (less-defined and separated), weak spots/noise (e.g., "ghost spots"), running differences between gels (e.g., protein migrates to different positions on different gels), unmatched/undetected spots, leading to missing values, mismatched spots , errors in quantification (several distinct spots may be erroneously detected as a single spot by the software and parts of a spot may be excluded from quantification), and differences in software algorithms and therefore analysis tendencies.

Generated picking lists can be used for the automated in-gel digestion of protein spots, and subsequent identification of the proteins by mass spectrometry.

BOTTOM-UP PROTEOMICS

Bottom-up proteomics is a common method to identify proteins and characterize their amino acid sequences and post-translational modifications by proteolytic digestion of proteins prior to analysis by mass spectrometry. The major alternative workflow used in proteomics is called top-down proteomics where intact proteins are purified prior to digestion and/or fragmentation either within the mass spectrometer or by 2D electrophoresis. Essentially, bottom-up proteomics is a relatively simple and reliable means of determining the protein make-up of a given sample of cells, tissues, etc.

In bottom-up proteomics, the crude protein extract is enzymatically digested, followed by one or more dimensions of separation of the peptides by liquid chromatography coupled to mass spectrometry, a technique known as shotgun proteomics. By comparing the masses of the proteolytic peptides or their tandem mass spectra with those predicted from a sequence database or annotated peptide spectral in a peptide spectral library, peptides can be identified and multiple peptide identifications assembled into a protein identification.

Advantages

For high throughput top-down methods, there is better front-end separation of peptides compared with proteins and higher sensitivity than the (non-gel) top-down methods.

Disadvantages

There is limited protein sequence coverage by identified peptides, loss of labile PTMs, and ambiguity of the origin for redundant peptide sequences. Recently the combination of bottom-up and top-down proteomics, so called middle-down proteomics, is receiving a lot of attention as this approach not only can be applied to the analysis of large protein fragments but also avoids redundant peptide sequences.

Shotgun Proteomics

Shotgun proteomics refers to the use of bottom-up proteomics techniques in identifying proteins in complex mixtures using a combination of high performance liquid chromatography combined with mass spectrometry. The name is derived from shotgun sequencing of DNA which is itself named after the rapidly expanding, quasi-random firing pattern of a shotgun. The most common method of shotgun proteomics starts with the proteins in the mixture being digested and the resulting peptides are separated by liquid chromatography. Tandem mass spectrometry is then used to identify the peptides.

Targeted proteomics using SRM and data-independent acquisition methods are often considered alternatives to shotgun proteomics in the field of bottom-up proteomics. While shotgun proteomics uses data-dependent selection of precursor ions to generate fragment ion scans, the aforementioned methods use a deterministic method for acquisition of fragment ion scans.

Shotgun proteomics arose from the difficulties of using previous technologies to separate complex mixtures. In 1975, two-dimensional polyacrylamide gel electrophoresis (2D-PAGE) was described by O'Farrell and Klose with the ability to resolve complex protein mixtures. The development of matrix-assisted laser desorption ionization (MALDI), electrospray ionization (ESI), and database searching continued to grow the field of proteomics. However these methods still had difficulty identifying and separating low-abundance proteins, aberrant proteins, and membrane proteins. Shotgun proteomics emerged as a method that could resolve even these proteins.

Advantages

Shotgun proteomics allows global protein identification as well as the ability to systematically profile dynamic proteomes. It also avoids the modest separation efficiency and poor mass spectral sensitivity associated with intact protein analysis.

Disadvantages

The dynamic exclusion filtering that is often used in shotgun proteomics maximizes the number of identified proteins at the expense of random sampling. This problem may be exacerbated by the undersampling inherent in shotgun proteomics.

Agilent 1200 HPLC

Quadrupole Time-of-Flight tandem
Mass Spectrometer (Q-TOF)

Workflow

Cells containing the protein complement desired are grown. Proteins are then extracted from the mixture and digested with a protease to produce a peptide mixture. The peptide mixture is then loaded directly onto a microcapillary column and the peptides are separated by hydrophobicity and charge. As the peptides elute from the column, they are ionized and separated by m/z in the first stage of tandem mass spectrometry. The selected ions undergo collision-induced dissociation or other process to induce fragmentation. The charged fragments are separated in the second stage of tandem mass spectrometry.

The "fingerprint" of each peptide's fragmentation mass spectrum is used to identify the protein from which they derive by searching against a sequence database with commercially available software (e.g. SEQUEST or MASCOT). Examples of sequence databases are the Genpept database or the PIR database. After the database search, each peptide-spectrum match (PSM) needs to be evaluated for validity. This analysis allows researchers to profile various biological systems.

Issues with Peptide Identification

Peptides that are degenerate (shared by two or more proteins in the database) may make it difficult to identify the protein to which they belong. Additionally, some proteome samples of vertebrates have a large number of paralogs. Finally, alternative splicing in higher eukaryotes can result in many identical protein subsequences.

Practical Applications

With the human genome sequenced, the next step is the verification and functional annotation of all predicted genes and their protein products. Shotgun proteomics can be used for functional classification or comparative analysis of these protein products. It can be used in projects ranging from large-scale whole proteome to focusing on a single protein family. It can be done in research labs or commercially.

Large-scale Analysis

One example of this is a study by Washburn, Wolters, & Yates in which they used shotgun proteomics on the proteome of a Saccharomyces cerevisiae strain grown to mid-log phase. They were able to detect and identify 1,484 proteins as well as identify proteins rarely seen in proteome analysis, including low-abundance proteins like transcription factors and protein kinases. They were also able to identify 131 proteins with three or more predicted transmembrane domains.

Protein Family

Vaisar et al. uses shotgun proteomics to implicate protease inhibition and complement activation in the antiinflammatory properties of high-density lipoprotein. In a study by Lee et al., higher expression level of hnRNP A2/B1 and Hsp90 were observed in human hepatoma HepG2 cells than in wild type cells. This led to a search for reported functional roles mediated in concert by both these multifunctional cellular chaperones.

THREADING

Protein threading, also known as fold recognition, is a method of protein modeling which is used to model those proteins which have the same fold as proteins of known structures, but do not have homologous proteins with known structure. It differs from the homology modeling method of structure prediction as it (protein threading) is used for proteins which do not have their homologous protein structures deposited in the Protein Data Bank (PDB), whereas homology modeling is used for those proteins which do. Threading works by using statistical knowledge of the relationship between the structures deposited in the PDB and the sequence of the protein which one wishes to model.

The prediction is made by "threading" (i.e. placing, aligning) each amino acid in the target sequence to a position in the template structure, and evaluating how well the target fits the template. After the best-fit template is selected, the structural model of the sequence is built based on the alignment with the chosen template. Protein threading is based on two basic observations: that the number of different folds in nature is fairly small (approximately 1300); and that 90% of the new structures submitted to the PDB in the past three years have similar structural folds to ones already in the PDB.

Classification of Protein Structure

The Structural Classification of Proteins (SCOP) database provides a detailed and comprehensive description of the structural and evolutionary relationships of known structure. Proteins are classified to reflect both structural and evolutionary relatedness. Many levels exist in the hierarchy, but the principal levels are family, superfamily and fold.

- Family (clear evolutionary relationship): Proteins clustered together into families are clearly evolutionarily related. Generally, this means that pairwise residue identities between the proteins are 30% and greater. However, in some cases similar functions and structures provide definitive evidence of common descent in the absence of high sequence identity; for example, many globins form a family though some members have sequence identities of only 15%.

- Superfamily (probable common evolutionary origin): Proteins that have low sequence identities, but whose structural and functional features suggest that a common evolutionary origin is probable, are placed together in superfamilies. For example, actin, the ATPase domain of the heat shock protein, and hexakinase together form a superfamily.

- Fold (major structural similarity): Proteins are defined as having a common fold if they have the same major secondary structures in the same arrangement and with the same topological connections. Different proteins with the same fold often have peripheral elements of secondary structure and turn regions that differ in size and conformation. In some cases, these differing peripheral regions may comprise half the structure. Proteins placed together in the same fold category may not have a common evolutionary origin: the structural similarities could arise just from the physics and chemistry of proteins favoring certain packing arrangements and chain topologies.

Method

A general paradigm of protein threading consists of the following four steps:

- The construction of a structure template database: Select protein structures from the protein structure databases as structural templates. This generally involves selecting protein structures from databases such as PDB, FSSP, SCOP, or CATH, after removing protein structures with high sequence similarities.

- The design of the scoring function: Design a good scoring function to measure the fitness between target sequences and templates based on the knowledge of the known relationships between the structures and the sequences. A good scoring function should contain mutation potential, environment fitness potential, pairwise potential, secondary structure compatibilities, and gap penalties. The quality of the energy function is closely related to the prediction accuracy, especially the alignment accuracy.

- Threading alignment: Align the target sequence with each of the structure templates by optimizing the designed scoring function. This step is one of the major tasks of all threading-based structure prediction programs that take into account the pairwise contact potential; otherwise, a dynamic programming algorithm can fulfill it.

- Threading prediction: Select the threading alignment that is statistically most probable as the threading prediction. Then construct a structure model for the target by placing the backbone atoms of the target sequence at their aligned backbone positions of the selected structural template.

Comparison with Homology Modeling

Homology modeling and protein threading are both template-based methods and there is no rigorous boundary between them in terms of prediction techniques. But the protein structures of their targets are different. Homology modeling is for those targets which have homologous proteins with known structure (usually/maybe of same family), while protein threading is for those targets with only fold-level homology found. In other words, homology modeling is for "easier" targets and protein threading is for "harder" targets.

Homology modeling treats the template in an alignment as a sequence, and only sequence homology is used for prediction. Protein threading treats the template in an alignment as a structure, and both sequence and structure information extracted from the alignment are used for prediction. When there is no significant homology found, protein threading can make a prediction based on the structure information. That also explains why protein threading may be more effective than homology modeling in many cases.

In practice, when the sequence identity in a sequence sequence alignment is low (i.e. <25%), homology modeling may not produce a significant prediction. In this case, if there is distant homology found for the target, protein threading can generate a good prediction.

More about Threading

Fold recognition methods can be broadly divided into two types: *1*, those that derive a 1-D profile for each structure in the fold library and align the target sequence to these profiles; and *2*, those that consider the full 3-D structure of the protein template. A simple example of a profile representation would be to take each amino acid in the structure and simply label it according to whether it is buried in the core of the protein or exposed on the surface. More elaborate profiles might take into account the local secondary structure (e.g. whether the amino acid is part of an alpha helix) or even evolutionary information (how conserved the amino acid is). In the 3-D representation, the structure is modeled as a set of inter-atomic distances, i.e. the distances are calculated between some or all of the atom pairs in the structure. This is a much richer and far more flexible description of the structure, but is much harder to use in calculating an alignment. The profile-based fold recognition approach was first described by Bowie, Lüthy and David Eisenberg in 1991. The term *threading* was first coined by David Jones, William R. Taylor and Janet Thornton in 1992, and originally referred specifically to the use of a full 3-D structure atomic representation of the protein template in fold recognition. Today, the terms threading and fold recognition are frequently (though somewhat incorrectly) used interchangeably.

Fold recognition methods are widely used and effective because it is believed that there are a strictly limited number of different protein folds in nature, mostly as a result of evolution but also due to constraints imposed by the basic physics and chemistry of polypeptide chains. There is, therefore, a good chance (currently 70-80%) that a protein which has a similar fold to the target protein has already been studied by X-ray crystallography or nuclear magnetic resonance (NMR) spectroscopy and can be found in the PDB. Currently there are nearly 1300 different protein folds known, but new folds are still being discovered every year due in significant part to the ongoing structural genomics projects.

Many different algorithms have been proposed for finding the correct threading of a sequence onto a structure, though many make use of dynamic programming in some form. For full 3-D threading, the problem of identifying the best alignment is very difficult (it is an NP-hard problem for some models of threading). Researchers have made use of many combinatorial optimization methods such as Conditional random fields, simulated annealing, branch and bound and linear programming, searching to arrive at heuristic solutions. It is interesting to compare threading methods to methods which attempt to align two protein structures (protein structural alignment), and indeed many of the same algorithms have been applied to both problems.

Protein Threading Software

- HHpred is a popular threading server which runs HHsearch, a widely used software for remote homology detection based on pairwise comparison of hidden Markov models.

- RAPTOR (software) is an integer programming based protein threading software. It has been replaced by a new protein threading program RaptorX / software for protein modeling and analysis, which employs probabilistic graphical models and statistical inference to both single template and multi-template based protein threading. RaptorX significantly outperforms RAPTOR and is especially good at aligning proteins with sparse sequence profile. The RaptorX server is free to public.

- Phyre is a popular threading server combining HHsearch with *ab initio* and multiple-template modelling.

- MUSTER is a standard threading algorithm based on dynamic programming and sequence profile-profile alignment. It also combines multiple structural resources to assist the sequence profile alignment.

- SPARKS X is a probabilistic-based sequence-to-structure matching between predicted one-dimensional structural properties of query and corresponding native properties of templates.

- BioShell is a threading algorithm using optimized profile-to-profile dynamic programming algorithm combined with predicted secondary structure.

PROTEIN MICROARRAY

A protein microarray (or protein chip) is a high-throughput method used to track the interactions and activities of proteins, and to determine their function, and determining function on a large scale. Its main advantage lies in the fact that large numbers of proteins can be tracked in parallel. The chip consists of a support surface such as a glass slide, nitrocellulose membrane, bead, or microtitre plate, to which an array of capture proteins is bound. Probe molecules, typically labeled with a fluorescent dye, are added to the array. Any reaction between the probe and the immobilised protein emits a fluorescent signal that is read by a laser scanner. Protein microarrays are rapid, automated, economical, and highly sensitive, consuming small quantities of samples and reagents. The concept and methodology of protein microarrays was first introduced and illustrated in antibody microarrays in 1983 in a scientific publication and a series of patents. The high-throughput technology behind the protein microarray was relatively easy to develop since it is based on the technology developed for DNA microarrays, which have become the most widely used microarrays.

Motivation for Development

Protein microarrays were developed due to the limitations of using DNA microarrays for determining gene expression levels in proteomics. The quantity of mRNA in the cell often doesn't reflect the expression levels of the proteins they correspond to. Since it is usually the protein, rather than the mRNA, that has the functional role in cell response, a novel approach was needed. Additionally post-translational modifications, which are often critical for determining protein function, are not visible on DNA microarrays. Protein microarrays replace traditional proteomics techniques such as 2D gel electrophoresis or chromatography, which were time consuming, labor-intensive and ill-suited for the analysis of low abundant proteins.

Making the Array

The proteins are arrayed onto a solid surface such as microscope slides, membranes, beads or microtitre plates. The function of this surface is to provide a support onto which proteins can be immobilized. It should demonstrate maximal binding properties, whilst maintaining the protein in its native conformation so that its binding ability is retained. Microscope slides made of glass or silicon are a popular choice since they are compatible with the easily obtained robotic arrayers and laser scanners that have been developed for DNA microarray technology. Nitrocellulose film slides are broadly accepted as the highest protein binding substrate for protein microarray applications.

The chosen solid surface is then covered with a coating that must serve the simultaneous functions of immobilising the protein, preventing its denaturation, orienting it in the appropriate direction so that its binding sites are accessible, and providing a hydrophilic environment in which the binding reaction can occur. It also needs to display minimal non-specific binding in order to minimize background noise in the detection systems. Furthermore, it needs to be compatible with different detection systems. Immobilising agents include layers of aluminium or gold, hydrophilic polymers, and polyacrylamide gels, or treatment with amines, aldehyde or epoxy. Thin-film technologies like physical vapour deposition (PVD) and chemical vapour deposition (CVD) are employed to apply the coating to the support surface.

An aqueous environment is essential at all stages of array manufacture and operation to prevent protein denaturation. Therefore, sample buffers contain a high percent of glycerol (to lower the freezing point), and the humidity of the manufacturing environment is carefully regulated. Microwells have the dual advantage of providing an aqueous environment while preventing cross-contamination between samples.

In the most common type of protein array, robots place large numbers of proteins or their ligands onto a coated solid support in a pre-defined pattern. This is known as robotic contact printing or robotic spotting. Another fabrication method is ink-jetting, a drop-on-demand, non-contact method of dispersing the protein polymers onto the solid surface in the desired pattern. Piezoelectric spotting is a similar method to ink-jet printing. The printhead moves across the array, and at each spot uses electric stimulation to deliver the protein molecules onto the surface via tiny jets. This is also a non-contact process. Photolithography is a fourth method of arraying the proteins onto the surface. Light is used in association with photomasks, opaque plates with holes or transparencies that allow light to shine through in a defined pattern. A series of chemical treatments then enables deposition of the protein in the desired pattern upon the material underneath the photomask.

The capture molecules arrayed on the solid surface may be antibodies, antigens, aptamers (nucleic acid-based ligands), affibodies (small molecules engineered to mimic monoclonal antibodies), or full length proteins. Sources of such proteins include cell-based expression systems for recombinant proteins, purification from natural sources, production in vitro by cell-free translation systems, and synthetic methods for peptides. Many of these methods can be automated for high throughput production but care must be taken to avoid conditions of synthesis or extraction that result in a denatured protein which, since it no longer recognizes its binding partner, renders the array useless.

Proteins are highly sensitive to changes in their microenvironment. This presents a challenge in maintaining protein arrays in a stable condition over extended periods of time. In situ methods — invented and published by Mingyue He and Michael Taussig in 2001 — involve on-chip synthesis of proteins as and when required, directly from the DNA using cell-free protein expression systems. Since DNA is a highly stable molecule it does not deteriorate over time and is therefore suited to long-term storage. This approach is also advantageous in that it circumvents the laborious and often costly processes of separate protein purification and DNA cloning, since proteins are made and immobilised simultaneously in a single step on the chip surface. Examples of in situ techniques are PISA (protein in situ array), NAPPA (nucleic acid programmable protein array) and DAPA (DNA array to protein array).

Types of Arrays

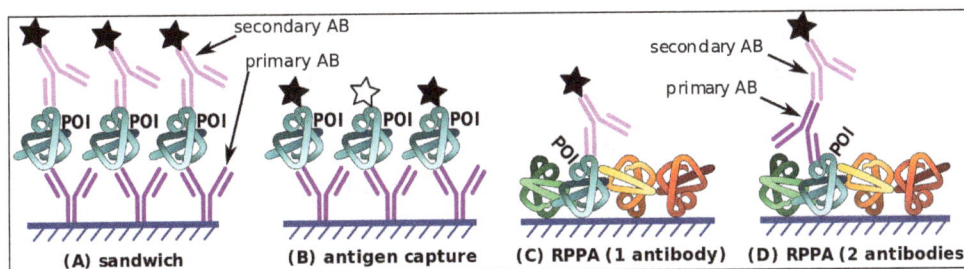

Types of protein arrays.

There are three types of protein microarrays that are currently used to study the biochemical activities of proteins.

Analytical microarrays are also known as capture arrays. In this technique, a library of antibodies, aptamers or affibodies is arrayed on the support surface. These are used as capture molecules since each binds specifically to a particular protein. The array is probed with a complex protein solution such as a cell lysate. Analysis of the resulting binding reactions using various detection systems can provide information about expression levels of particular proteins in the sample as well as measurements of binding affinities and specificities. This type of microarray is especially useful in comparing protein expression in different solutions. For instance the response of the cells to a particular factor can be identified by comparing the lysates of cells treated with specific substances or grown under certain conditions with the lysates of control cells. Another application is in the identification and profiling of diseased tissues.

Reverse phase protein microarray (RPPA) involve complex samples, such as tissue lysates. Cells are isolated from various tissues of interest and are lysed. The lysate is arrayed onto the microarray and probed with antibodies against the target protein of interest. These antibodies are typically detected with chemiluminescent, fluorescent or colorimetric assays. Reference peptides are printed on the slides to allow for protein quantification of the sample lysates. RPAs allow for the determination of the presence of altered proteins or other agents that may be the result of disease. Specifically, post-translational modifications, which are typically altered as a result of disease can be detected using RPAs.

Functional Protein Microarrays

Functional protein microarrays (also known as target protein arrays) are constructed by immobilising large numbers of purified proteins and are used to identify protein–protein, protein–DNA, protein–RNA, protein–phospholipid, and protein–small-molecule interactions, to assay enzymatic activity and to detect antibodies and demonstrate their specificity. They differ from analytical arrays in that functional protein arrays are composed of arrays containing full-length functional proteins or protein domains. These protein chips are used to study the biochemical activities of the entire proteome in a single experiment.

The key element in any functional protein microarray-based assay is the arrayed proteins must retain their native structure, such that meaningful functional interactions can take place on the array surface. The advantages of controlling the precise mode of surface attachment through use of an appropriate affinity tag are that the immobilised proteins will have a homogeneous orientation resulting in a higher specific activity and higher signal-to-noise ratio in assays, with less interference from non-specific interactions.

Detection

Protein array detection methods must give a high signal and a low background. The most common and widely used method for detection is fluorescence labeling which is highly sensitive, safe and compatible with readily available microarray laser scanners. Other labels can be used, such as affinity, photochemical or radioisotope tags. These labels are attached to the probe itself and can interfere with the probe-target protein reaction. Therefore, a number of label free detection methods are available, such as surface plasmon resonance (SPR), carbon nanotubes, carbon nanowire sensors (where detection occurs via changes in conductance) and microelectromechanical system (MEMS) cantilevers. All these label free detection methods are relatively new and are not yet suitable for high-throughput protein interaction detection; however, they do offer much promise for the future.

Protein quantitation on nitrocellulose coated glass slides can use near-IR fluorescent detection. This limits interferences due to auto-fluorescence of the nitrocellulose at the UV wavelengths used for standard fluorescent detection probes.

Applications

There are five major areas where protein arrays are being applied: diagnostics, proteomics, protein functional analysis, antibody characterization, and treatment development.

Diagnostics involves the detection of antigens and antibodies in blood samples; the profiling of sera to discover new disease biomarkers; the monitoring of disease states and responses to therapy in personalized medicine; the monitoring of environment and food. Digital bioassay is an example of using protein microarray for diagnostic purposes. In this technology, an array of microwells on a glass/polymer chip are seeded with magnetic beads (coated with fluorescent tagged antibodies), subjected to targeted antigens and then characterised by a microscope through counting fluorescing wells. A cost-effective fabrication platform (using OSTE polymers) for such microwell arrays has been recently demonstrated and the bio-assay model system has been successfully characterised.

Proteomics pertains to protein expression profiling i.e. which proteins are expressed in the lysate of a particular cell.

Protein functional analysis is the identification of protein–protein interactions (e.g. identification of members of a protein complex), protein–phospholipid interactions, small molecule targets, enzymatic substrates (particularly the substrates of kinases) and receptor ligands.

Antibody characterization is characterizing cross-reactivity, specificity and mapping epitopes.

Treatment development involves the development of antigen-specific therapies for autoimmunity, cancer and allergies; the identification of small molecule targets that could potentially be used as new drugs.

Challenges

Despite the considerable investments made by several companies, proteins chips have yet to flood the market. Manufacturers have found that proteins are actually quite difficult to handle. Production of reliable, consistent, high-throughput proteins that are correctly folded and functional is fraught with difficulties as they often result in low-yield of proteins due to decreased solubility and formation of inclusion bodies. A protein chip requires a lot more steps in its creation than does a DNA chip.

There are a number of approaches to this problem which differ fundamentally according to whether the proteins are immobilised through non-specific, poorly defined interactions, or through a specific set of known interactions. The former approach is attractive in its simplicity and is compatible with purified proteins derived from native or recombinant sources but suffers from a number of risks. Most notable amongst these relate to the uncontrolled nature of the interactions between each protein and the surface; at best, this might give rise to a heterogeneous population of proteins in which active sites are sometimes occluded by the surface; at worst, it might destroy activity altogether due to partial or complete surface-mediated unfolding of the immobilised protein.

Challenges include: 1) finding a surface and a method of attachment that allows the proteins to maintain their secondary or tertiary structure and thus their biological activity and their interactions with other molecules, 2) producing an array with a long shelf life so that the proteins on the chip do not denature over a short time, 3) identifying and isolating antibodies or other capture molecules against every protein in the human genome, 4) quantifying the levels of bound protein while assuring sensitivity and avoiding background noise, 5) extracting the detected protein from the chip in order to further analyze it, 6) reducing non-specific binding by the capture agents, 7) the capacity of the chip must be sufficient to allow as complete a representation of the proteome to be visualized as possible; abundant proteins overwhelm the detection of less abundant proteins such as signaling molecules and receptors, which are generally of more therapeutic interest.

LOOP MODELING

Loop modeling is a problem in protein structure prediction requiring the prediction of the conformations of loop regions in proteins with or without the use of a structural template. Computer programs that solve these problems have been used to research a broad range of scientific topics from ADP to breast cancer. Because protein function is determined by its shape and the physiochemical properties of its exposed surface, it is important to create an accurate model for protein/ligand interaction studies. The problem arises often in homology modeling, where the tertiary structure of an amino acid sequence is predicted based on a sequence alignment to a *template*, or a second sequence whose structure is known. Because loops have highly variable sequences even within a given structural motif or protein fold, they often correspond to unaligned regions in sequence alignments; they also tend to be located at the solvent-exposed surface of globular proteins and thus are more conformationally flexible. Consequently, they often cannot be modeled using standard homology modeling techniques. More constrained versions of loop modeling are also used in the data fitting stages of solving a protein structure by X-ray crystallography, because loops can correspond to regions of low electron density and are therefore difficult to resolve.

Regions of a structural model that are predicted by non-template-based loop modeling tend to be much less accurate than regions that are predicted using template-based techniques. The extent of the inaccuracy increases with the number of amino acids in the loop. The loop amino acids' side chains dihedral angles are often approximated from a rotamer library, but can worsen the inaccuracy of side chain packing in the overall model. Andrej Sali's homology modeling suite MODELLER includes a facility explicitly designed for loop modeling by a satisfaction of spatial restraints method. All methods require an upload of the PDB file and some require the specification of the loop location.

Short Loops

In general, the most accurate predictions are for loops of fewer than 8 amino acids. Extremely short loops of three residues can be determined from geometry alone, provided that the bond lengths and bond angles are specified. Slightly longer loops are often determined from a "spare parts" approach, in which loops of similar length are taken from known crystal structures and adapted to the geometry of the flanking segments. In some methods, the bond lengths and angles of the loop region are allowed to vary, in order to obtain a better fit; in other cases, the constraints of the flanking segments may be varied to find more "protein-like" loop conformations. The accuracy of such short loops may be almost as accurate as that of the homology model upon which it is based. It should also be considered that the loops in proteins may not be well-structured and therefore have no one conformation that could be predicted; NMR experiments indicate that solvent-exposed loops are "floppy" and adopt many conformations, while the loop conformations seen by X-ray crystallography may merely reflect crystal packing interactions, or the stabilizing influence of crystallization co-solvents.

Template based Techniques

As mentioned above homology-based methods use a database to align the target protein gap with a known template protein. A database of known structures is searched for a loop that fits the gap of interest by similarity of sequence and stems (the edges of the gap created by the unknown loop structure). The success of this method largely depends on the quality of that alignment. Since the loop is the least conserved portion of a protein's structure, the homology-based method cannot always find a known template that aligns with the target sequence. Fortunately, the template databases are always adding new templates so the problem of not being able to find an alignment is becoming less of an issue. Some programs that use this method are SuperLooper and FREAD.

Non-template based Techniques

Otherwise known as an *ab initio* method, non-template based approaches use a statistical model to fill in the gaps created by the unknown loop structure. Some of these programs include MODELLER, Loopy, and RAPPER; but each of these programs approaches the problem in a different manner. For example, Loopy uses samples of torsion angle pairs to generate the initial loop structure then it revises this structure to maintain a realistic shape and closure, while RAPPER builds from one end of the gap to the other by extending the stem with different sampled angles until the gap is closed. Yet another method is the "divide and conquer" approach. This involves subdividing the loop into 2 segments and then repeatedly dividing and transforming each segment until the loop is small enough to be solved. Even with all these methods non-template based approaches are most accurate up to 12 residues (amino acids within the loop).

There are three problems that arise when using a non-template based technique. First, there are constraints that limit the possibilities for local region modeling. One such constraint is that loop termini are required to end at the correct anchor position. Also, the Ramachandran space cannot contain a backbone of dihedral angles. Second, a modeling program has to use a set procedure. Some programs use the "spare parts" approach as mentioned above. Other programs use a *de novo* approach that samples sterically feasible loop conformations and selects the best one. Third, determining the best model means that a scoring method must be created to compare the various conformations.

PROTEOSTASIS

Proteostasis, a portmanteau of the words protein and homeostasis, is the concept that there are competing and integrated biological pathways within cells that control the biogenesis, folding, trafficking and degradation of proteins present within and outside the cell. The concept of proteostasis maintenance is central to understanding the cause of diseases associated with excessive protein misfolding and degradation leading to loss-of-function phenotypes, as well as aggregation-associated degenerative disorders. Therefore, adapting proteostasis should enable the restoration of proteostasis once its loss leads to pathology. Cellular proteostasis is key to ensuring successful development, healthy aging, resistance to environmental stresses, and to minimize homeostasis perturbations by pathogens such as viruses. Mechanisms by which proteostasis is ensured include regulated protein translation, chaperone assisted protein folding and protein degradation pathways. Adjusting each of these mechanisms to the demand for proteins is essential to maintain all cellular functions relying on a correctly folded proteome.

Mechanisms of Proteostasis

The Roles of the Ribosome in Proteostasis

One of the first points of regulation for proteostasis is during translation. This is accomplished via the structure of the ribosome, a complex central to translation. These two characteristics shape the way the protein folds and influences the proteins future interactions. The synthesis of a new peptide chain using the ribosome is very slow and the ribosome can even be stalled when it encounters a rare codon, a codon found at low concentrations in the cell. These pauses provide an opportunity for an individual protein domain to have the necessary time to become folded before the production of following domains. This facilitates the correct folding of multi-domain proteins. The newly synthesized peptide chain exits the ribosome into the cellular environment through the narrow ribosome exit channel (width: 10Å to 20Å, length 80Å). Due to space restriction in the exit channel the nascent chain already forms secondary and limited tertiary structures. For example, an alpha helix is one such structural property that is commonly induced in this exit channel. At the same time the exit channel also prevents premature folding by impeding large scale interactions within the peptide chain which would require more space.

Molecular Chaperones and Post-translational Maintenance in Proteostasis

In order to maintain protein homeostasis post-translationally, the cell makes use of molecular chaperones sometimes including chaperonins, which aid in the assembly or disassembly of proteins. They recognize exposed segments of hydrophobic amino acids in the nascent peptide chain and then work to promote the proper formation of noncovalent interactions that lead to the desired folded state. Chaperones begin to assist in protein folding as soon as a nascent chain longer than 60 amino acids emerges from the ribosome exit channel. One of the most studied ribosome binding chaperones is trigger factor. Trigger factor works to stabilize the peptide, promotes its folding, prevents aggregation, and promotes refolding of denatured model substrates. Trigger factor not only directly works to properly fold the protein but also recruits other chaperones to the ribosome, such as Hsp70. Hsp70 surrounds an unfolded peptide chain, thereby preventing aggregation and promoting folding.

Chaperonins are a special class of chaperones that promote native state folding by cyclically encapsulating the peptide chain. Chaperonins are divided into two groups. Group 1 chaperonins are commonly found in bacteria, chloroplasts, and mitochondria. Group 2 chaperonins are found in both the cytosol of eukaryotic cells as well as in archaea. Group 2 chaperonins also contain an additional helical component which acts as a lid for the cylindrical protein chamber, unlike Group 1 which instead relies on an extra cochaperone to act as a lid. All chaperonins exhibit two states (open and closed), between which they can cycle. This cycling process is important during the folding of an individual polypeptide chain as it helps to avoid undesired interactions as well as to prevent the peptide from entering into kinetically trapped states.

Regulating Proteostasis by Protein Degradation

The third component of the proteostasis network is the protein degradation machinery. Protein degradation occurs in proteostasis when the cellular signals indicate the need to decrease overall cellular protein levels. The effects of protein degradation can be local, with the cell only experiencing effects from the loss of the degraded protein itself or widespread, with the entire protein landscape changing due to loss of other proteins' interactions with the degraded protein. Multiple substrates are targets for proteostatic degradation. These degradable substrates include nonfunctional protein fragments produced from ribosomal stalling during translation, misfolded or unfolded proteins, aggregated proteins, and proteins that are no longer needed to carry out cellular function. Several different pathways exist for carrying out these degradation processes. When proteins are determined to be unfolded or misfolded, they are typically degraded via the unfolded protein response (UPR) or endoplasmic-reticulum-associated protein degradation (ERAD). Substrates that are unfolded, misfolded, or no longer required for cellular function can also be ubiquitin tagged for degradation by ATP dependent proteases, such as the proteasome in eukaryotes or ClpXP in prokaryotes. Autophagy, or self engulfment, lysosomal targeting, and phagocytosis (engulfment of waste products by other cells) can also be used as proteostatic degradation mechanisms.

Signaling Events in Proteostasis

Protein misfolding is detected by mechanisms that are specific for the cellular compartment in which they occur. Distinct surveillance mechanisms that respond unfolded protein have been characterized in the cytoplasm, ER and mitochondria. This response acts locally in a cell autonomous fashion but can also extend to intercellular signaling to protect the organism from anticipated proteotoxic stress.

Cell-autonomous Stress Responses

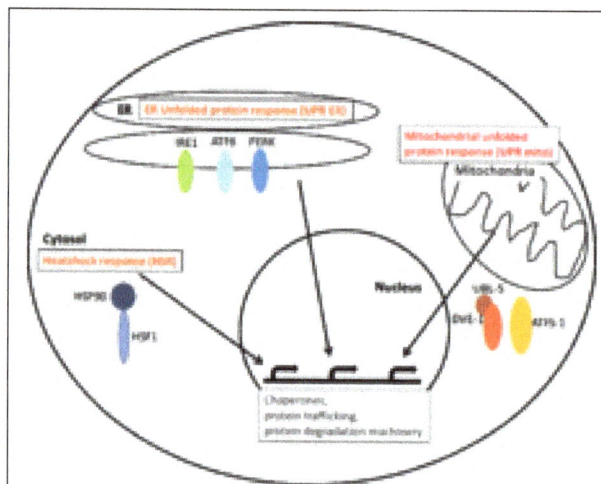

Proteostasis stress signaling response.

Cellular stress response pathways detect and alleviate proteotoxic stress which is triggered by imbalances in proteostasis. The cell-autonomous regulation occurs through direct detection of misfolded proteins or inhibition of pathway activation by sequestering activating components in response to heat shock. Cellular responses to this stress signaling include transcriptional activation of chaperone expression, increased efficiency in protein trafficking and protein degradation and translational reduction.

Cytosolic Heat Shock Response

The cytosolic HSR is mainly mediated by the transcription factor family HSF (heat shock family). HSF is constitutively bound by Hsp90. Upon a proteotoxic stimulus Hsp90 is recruited away from HSF which can then bind to heat response elements in the DNA and upregulate gene expression of proteins involved in the maintenance of proteostasis.

ER Unfolded Protein Response

The unfolded protein response in the endoplasmatic reticulum (ER) is activated by imbalances of unfolded proteins inside the ER and the proteins mediating protein homeostasis. Different "detectors" - such as IRE1, ATF6 and PERK - can recognize misfolded proteins in the ER and mediate transcriptional responses which help alleviate the effects of ER stress.

Mitochondrial Unfolded Protein Response

The mitochondrial unfolded protein response detects imbalances in protein stoichiometry of mitochondrial proteins and misfolded proteins. The expression of mitochondrial chaperones is upregulated by the activation of the transcription factors ATF-1 and/or DVE-1 with UBL-5.

Systemic Stress Signaling

Stress responses can also be triggered in a non-cell autonomous fashion by intercellular communication. The stress that is sensed in one tissue could thereby be communicated to other tissues to protect the proteome of the organism or to regulate proteostasis systemically. Cell non-autonomous activation can occur for all three stress responses.

Work on the model organism *C. elegans* has shown that neurons play a role in this intercellular communication of cytosolic HSR. Stress induced in the neurons of the worm can in the long run protect other tissues such as muscle and intestinal cells from chronic proteotoxicity. Similarly ER and mitochondrial UPR in neurons are relayed to intestinal cells . These systemic responses have been implicated in mediating not only systemic proteostasis but also influence organismal aging.

Proteostasis and Diseases of Protein Folding

Dysfunction in proteostasis can arise from errors in or misregulation of protein folding. The classic examples are missense mutations and deletions that change the thermodynamic and kinetic parameters for the protein folding process. These mutations are often inherited and range in phenotypic severity from having no noticeable effect to embryonic lethality. Disease develops when these mutations render a protein significantly more susceptible to misfolding, aggregation, and degradation. If these effects only alter the mutated protein, the negative consequences will only be local loss of function. However, if these mutations occur in a chaperone or a protein that interacts with many other proteins, dramatic global alterations in the proteostasis boundary will occur. Examples of diseases resulting from proteostatic changes from errors in protein folding include cystic fibrosis, Huntington's disease, Alzheimer's disease, lysosomal storage disorders, and others.

Role of Model Systems in the Elucidation of Protein-misfolding Diseases

Small animal model systems have been and continue to be instrumental in the identification of functional mechanisms that safeguard proteostasis. Model systems of diverse misfolding-prone disease proteins have so far revealed numerous chaperone and co-chaperone modifiers of proteotoxicity.

Proteostasis and Cancer

The unregulated cell division that marks cancer development requires increased protein synthesis for cancer cell function and survival. This increased protein synthesis is typically seen in proteins that modulate cell metabolism and growth processes. Cancer cells are sometimes susceptible to drugs that inhibit chaperones and disrupt proteostasis, such as Hsp90 inhibitors or proteasome inhibitors. Furthermore, cancer cells tend to produce misfolded proteins, which are removed mainly by proteolysis. Inhibitors of proteolysis allow accumulation of both misfolded protein aggregates, as well as apoptosis signaling proteins in cancer cells. This can change the sensitivity of cancer cells to antineoplastic drugs; cancer cells either die at a lower drug concentration, or survive, depending on the type of proteins that accumulate, and the function these proteins have. Proteasome inhibitor bortezomib was the first drug of this type to receive approval for treatment of multiple myeloma.

Proteostasis and Obesity

A hallmark of cellular proteostatic networks is their ability to adapt to stress via protein regulation. Metabolic disease, such as that associated with obesity, alters the ability of cellular proteostasis networks adapt to stress, often with detrimental health effects. For example, when insulin production exceeds the cell's insulin secretion capacity, proteostatic collapse occurs and chaperone production is severely impaired. This disruption leads to the disease symptoms exhibited in individuals with diabetes.

Proteostasis and Aging

Over time, the proteostasis network becomes burdened with proteins modified by reactive oxygen species and metabolites that induce oxidative damage. These byproducts can react with cellular proteins to cause misfolding and aggregation (especially in nondividing cells like neurons). This risk is particularly high for intrinsically disordered proteins. The IGFR-1 pathway has been shown in *C. elegans* to protect against these harmful aggregates, and some experimental work has suggested that upregulation of insulin growth factor receptor 1 (IGFR-1) may stabilize proteostatic network and prevent detrimental effects of aging. Expression of the chaperome, the ensemble of chaperones and co-chaperones that interact in a complex network of molecular folding machines to regulate proteome function, is dramatically repressed in human aging brains and in the brains of patients with neurodegenerative diseases. Functional assays in *C. elegans* and human cells have identified a conserved chaperome sub-network of 16 chaperone genes, corresponding to 28 human orthologs as a proteostasis safeguard in aging and age-onset neurodegenerative disease.

Pharmacologic Intervention in Proteostasis

There are two main approaches that have been used for therapeutic development targeting the proteostatic network: pharmacologic chaperones and proteostasis regulators. The principle behind designing pharmacologic chaperones for intervention in diseases of proteostasis is to design small molecules that stabilize proteins exhibiting borderline stability. Previously, this approach has been used to target and stabilize G-protein coupled receptors, neurotransmitter receptors, glycosidases, lysosomal storage proteins, and the mutant CFTR protein that causes cystic fibrosis and transthyretin, which can misfiled and aggregate leading to amyloidoses. Vertex Pharmaceuticals and Pfizer sell regulatory agency approved pharmacologic chaperones for ameliorating cystic fibrosis and the transthyretin amyloidoses, respectively. Amicus sells a regulatory agency approved pharmacologic chaperone for Fabry disease–a lysosomal storage disease.

The principle behind proteostasis regulators is different, these molecules alter the biology of protein folding and degradation by altering the stoichiometry of the proteostasis network components in a given sub cellular compartment. For example, some proteostasis regulators initiate stress responsive signaling, such as the unfolded protein response, which transcriptionally reprograms the endoplasmic reticulum proteostasis network. It has been suggested that this approach could even be applied prophylactically, such as upregulating certain protective pathways before experiencing an anticipated severe cellular stress. One theoretical mechanism for this approach includes upregulating the heat shock response response to rescue proteins from degradation during cellular stress.

References

- Shevchenko A, Tomas H, Havlis J, Olsen JV, Mann M (2006). "In-gel digestion for mass spectrometric characterization of proteins and proteomes". Nature Protocols. 1 (6): 2856–60. Doi:10.1038/nprot.2006.468. PMID 17406544

- What-is-protein-purification, affinity-chromatography, protein-analysis-guide, references, home: thermofisher.com, Retrieved 25 April, 2019

- Mikkelsen, Susan; Cortón, Eduardo (2004). Bioanalytical Chemistry. John Wiley & Sons, Inc. P. 224. ISBN 978-0-471-62386-1

- Protein-separation-techniques, publications-resources, default-source: meatscience.org, Retrieved 26 May 2019 "DNA Microarrays: Techniques". Arabidopsis.info. Archived from the original on August 28, 2008. Retrieved January 19,2013

- Overview-electrophoresis, pierce-protein-methods, protein-biology-resource-library, protein-biology-learning-center, protein-biology, life-science, home: thermofisher.com, Retrieved 27 June, 2019

- Bogosian G, Violand BN, Dorward-King EJ, Workman WE, Jung PE, Kane JF (January 1989). "Biosynthesis and incorporation into protein of norleucine by Escherichia coli". The Journal of Biological Chemistry. 264 (1): 531–9. PMID 2642478

- Protein-identification-methods-3: creative-proteomics.com, Retrieved 28 July, 2019

- Durbin, Kenneth Robert; Fornelli, Luca; Fellers, Ryan T.; Doubleday, Peter F.; Narita, Masashi; Kelleher, Neil L. (2016). "Quantitation and Identification of Thousands of Human Proteoforms Below 30 kda". Journal of Proteome Research. 15 (3): 976–982. Doi:10.1021/acs.jproteome.5b00997. PMC 4794255. PMID 26795204

- Overview-mass-spectrometry, pierce-protein-methods, protein-biology-resource-library, protein-biology-learning-center, protein-biology, life-science, home: thermofisher.com, Retrieved 28 July, 2019

Protein-Protein Interaction

The physical contacts of high specificity established between two or more protein molecules are referred to as protein–protein interactions. The main focus areas of protein-protein interaction are protein–protein interaction prediction and protein–protein interaction screening. This chapter has been carefully written to provide an easy understanding of these aspects associated with protein-protein interaction.

Proteins are the workhorses that facilitate most biological processes in a cell, including gene expression, cell growth, proliferation, nutrient uptake, morphology, motility, intercellular communication and apoptosis. But cells respond to a myriad of stimuli, and therefore protein expression is a dynamic process; the proteins that are used to complete specific tasks may not always be expressed or activated. Additionally, all cells are not equal, and many proteins are expressed in a cell type–dependent manner. These basic characteristics of proteins suggest a complexity that can be difficult to investigate, especially when trying to understand protein function in the proper biological context.

Critical aspects required to understand the function of a protein include:

- Protein sequence and structure: Used to discover motifs that predict protein function.

- Evolutionary history and conserved sequences: Identifies key regulatory residues.

- Expression profile: Reveals cell-type specificity and how expression is regulated.

- Post-translational modifications: Phosphorylation, acylation, glycosylation and ubiquitination suggest localization, activation and function.

- Interactions with other proteins: Function may be extrapolated by knowing the function of binding partners.

- Intracellular localization: May allude to the function of the protein.

Until the late 1990s, protein function analyses mainly focused on single proteins. However, because the majority of proteins interact with other proteins for proper function, they should be studied in the context of their interacting partners to fully understand their function. With the publication of the human genome and the development of the field of proteomics, understanding how proteins interact with each other and identifying biological networks has become vital to understanding how proteins function within the cell.

Types of Protein–protein Interactions

Protein interactions are fundamentally characterized as stable or transient, and both types of interactions can be either strong or weak. Stable interactions are those associated with proteins that are purified as multi-subunit complexes, and the subunits of these complexes can be identical or different. Hemoglobin and core RNA polymerase are examples of multi-subunit interactions that form stable complexes.

Transient interactions are expected to control the majority of cellular processes. As the name implies, transient interactions are temporary in nature and typically require a set of conditions that promote the interaction, such as phosphorylation, conformational changes or localization to discrete areas of the cell. Transient interactions can be strong or weak, and fast or slow. While in contact with their binding partners, transiently interacting proteins are involved in a wide range of cellular processes, including protein modification, transport, folding, signaling, apoptosis and cell cycling. The following example provides an illustration of protein interactions that regulate apoptotic and anti-apoptotic processes.

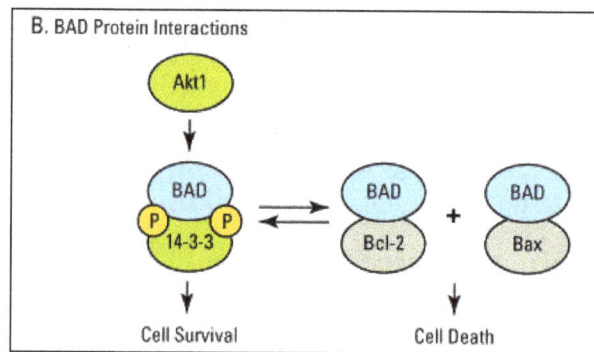

Heavy BAD protein–protein interaction. Panel A: Coomassie-stained SDS-PAGE gel of recombinant light and heavy BAD-GST-HA-6xHIS purified from HeLa IVT lysates (L), using glutathione resin (E1) and cobalt resin (E2) tandem affinity. The flow-through (FT) from each column is indicated. Panel B: Schematic of BAD phosphorylation and protein interactions during cell survival and cell death (i.e., apoptosis). Panel C: BAD protein sequence coverage showing identified Akt consensus phosphorylation sites (red box). Panel D: MS spectra of stable isotope-labeled BAD peptide HSSYPAGTEDDEGmGEEPSPFr.

Proteins bind to each other through a combination of hydrophobic bonding, van der Waals forces, and salt bridges at specific binding domains on each protein. These domains can be small binding clefts or large surfaces and can be just a few peptides long or span hundreds of amino acids. The strength of the binding is influenced by the size of the binding domain. One example of a common surface domain that facilitates stable protein–protein interactions is the leucine zipper, which consists of α-helices on each protein that bind to each other in a parallel fashion through the hydrophobic bonding of regularly-spaced leucine residues on each α-helix that project between the adjacent helical peptide chains. Because of the tight molecular packing, leucine zippers provide stable binding for multi-protein complexes, although all leucine zippers do not bind identically due to non-leucine amino acids in the α-helix that can reduce the molecular packing and therefore the strength of the interaction.

Two Src homology (SH) domains, SH2 and SH3, are examples of common transient binding domains that bind short peptide sequences and are commonly found in signaling proteins. The SH2 domain recognizes peptide sequences with phosphorylated tyrosine residues, which are often indicative of protein activation. SH2 domains play a key role in growth factor receptor signaling, during which ligand-mediated receptor phosphorylation at tyrosine residues recruits downstream effectors that recognize these residues via their SH2 domains. The SH3 domain usually recognizes proline-rich peptide sequences and is commonly used by kinases, phospholipases and GTPases to identify target proteins. Although both SH2 and SH3 domains generally bind to these motifs, specificity for distinct protein interactions is dictated by neighboring amino acid residues in the respective motif.

Biological Effects of Protein–protein Interactions

The result of two or more proteins that interact with a specific functional objective can be demonstrated in several different ways. The measurable effects of protein interactions have been outlined as follows:

- Alter the kinetic properties of enzymes, which may be the result of subtle changes in substrate binding or allosteric effects.

- Allow for substrate channeling by moving a substrate between domains or subunits, resulting ultimately in an intended end product.

- Create a new binding site, typically for small effector molecules.

- Inactivate or destroy a protein.

- Change the specificity of a protein for its substrate through the interaction with different binding partners, e.g., demonstrate a new function that neither protein can exhibit alone.

- Serve a regulatory role in either an upstream or a downstream event.

Common Methods to Analyze Protein–protein Interactions

Usually a combination of techniques is necessary to validate, characterize and confirm protein interactions. Previously unknown proteins may be discovered by their association with one or more proteins that are known. Protein interaction analysis may also uncover unique, unforeseen functional roles for well-known proteins. The discovery or verification of an interaction is the first step on the road to understanding where, how and under what conditions these proteins interact in vivo and the functional implications of these interactions.

While the various methods and approaches to studying protein–protein interactions are too numerous to describe here, the table below and the remainder of this topic focuses on common methods to analyze protein–protein interactions and the types of interactions that can be studies using each method. In summary, stable protein–protein interactions are easiest to isolate by physical methods like co-immunoprecipitation and pull-down assays because the protein complex does not disassemble over time. Weak or transient interactions can be identified using these methods by first covalently crosslinking the proteins to freeze the interaction during the co-IP or pull-down. Alternatively, crosslinking, along with label transfer and far–western blot analysis, can be performed independent of other methods to identify protein–protein interactions.

Common methods to analyze the various types of protein interactions:

Method	Protein–protein interactions
Co-immunoprecipitation (co-IP)	Stable or strong
Pull-down assay	Stable or strong
Crosslinking protein interaction analysis	Transient or weak
Label transfer protein interaction analysis	Transient or weak
Far–western blot analysis	Moderately stable

Co-immunoprecipitation

Co-immunoprecipitation (co-IP) is a popular technique for protein interaction discovery. Co-IP is conducted in essentially the same manner as an immunoprecipitation (IP) of a single protein, except that the target protein precipitated by the antibody, also called the "bait", is used to co-precipitate a binding partner/protein complex, or "prey", from a lysate. Essentially, the interacting protein is bound to the target antigen, which is bound by the antibody that is immobilized to the support. Immunoprecipitated proteins and their binding partners are commonly detected by sodium dodecyl sulfate–polyacrylamide gel electrophoresis (SDS-PAGE) and western blot analysis. The assumption that is usually made when associated proteins are co-precipitated is that these proteins are related to the function of the target antigen at the cellular level. This is only an assumption, however, that is subject to further verification.

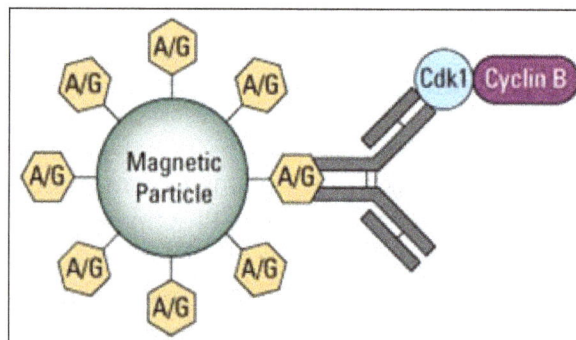

Co-immunoprecipitation of cyclin B and Cdk1.

The Thermo Scientific Pierce Protein A/G Magnetic Beads bind to Cdk1 antibody complexed with Cdk1. Cyclin B is bound to the Cdk1, and is captured along with its binding partner.

Pull-down Assays

Pull-down assays are similar in methodology to co-immunoprecipitation because of the use of beaded support to purify interacting proteins. The difference between these two approaches, though, is that while co-IP uses antibodies to capture protein complexes, pull-down assays use a "bait" protein to purify any proteins in a lysate that bind to the bait. Pull-down assays are ideal for studying strong or stable interactions or those for which no antibody is available for co-immunoprecipitation.

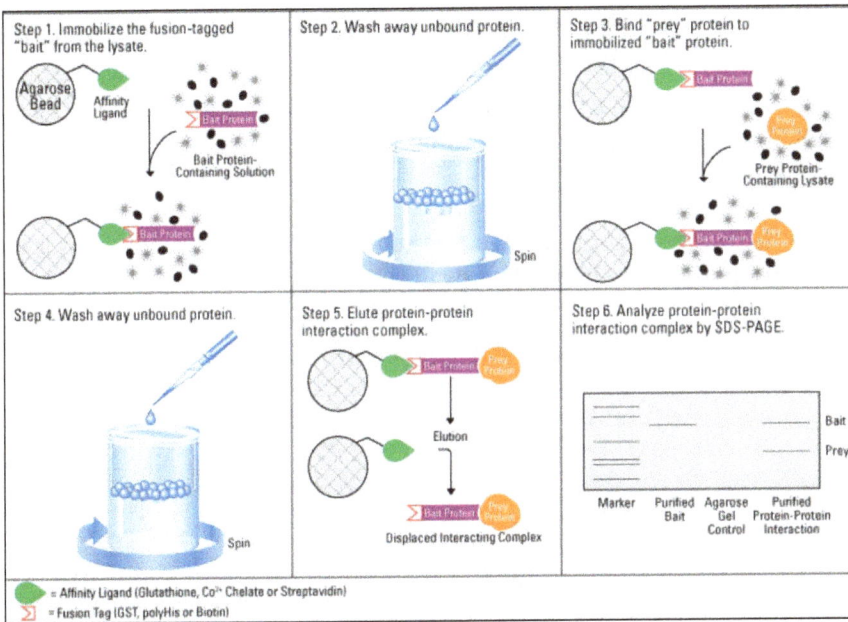

General schematic of a pull-down assay. A pull-down assay is a small-scale affinity purification technique similar to immunoprecipitation, except that the antibody is replaced by some other affinity system. In this case, the affinity system consists of a glutathione S-transferase (GST)–, polyHis- or streptavidin-tagged protein or binding domain that is captured by glutathione-, metal chelate (cobalt or nickel) – or biotin-coated agarose beads, respectively. The immobilized fusion-tagged protein acts as the "bait" to capture a putative binding partner (i.e., the "prey"). In a typical pull-down assay, the immobilized bait protein is incubated with a cell lysate, and after the prescribed washing steps, the complexes are selectively eluted using competitive analytes or low pH or reducing buffers for in-gel or western blot analysis.

Crosslinking Protein Interaction Analysis

Most protein–protein interactions are transient, occurring only briefly as part of a single cascade or other metabolic function within cells. Crosslinking interacting proteins is an approach to stabilize or permanently adjoin the components of interaction complexes. Once the components of an interaction are covalently crosslinked, other steps (e.g., cell lysis, affinity purification, electrophoresis or mass spectrometry) can be used to analyze the protein–protein interaction while maintaining the original interacting complex.

Homobifunctional, amine-reactive crosslinkers can be added to cells to crosslink potentially interacting proteins together, which can then be analyzed after lysis by western blotting. Crosslinkers can be membrane permeable, such as DSS, for crosslinking intracellular proteins, or they can be non–membrane permeable, such as BS3, for crosslinking cell-surface proteins. Furthermore, some crosslinkers can be cleaved by reducing agents, such as DSP or DTSSP, to reverse the crosslinks.

Alternatively, heterobifunctional crosslinkers that contain a photoactivatable group, such as SDA product or Sulfo-SDA, can be used to capture transient interactions that may occur, such as after a particular stimulus. Photoactivation can also be also be after metabolic labeling with photoactivatable amino acids such as L-Photo-Leucine or L-Photo-Methionine.

Crosslinking sites between proteins can be mapped by high precision using mass spectrometry, especially if a MS-cleavable crosslinker such as DSSO or DSBU is used.

Label Transfer Protein Interaction Analysis

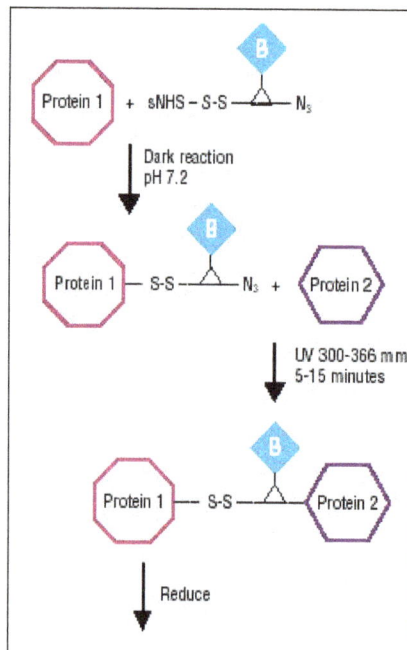

Experimental strategy for Sulfo-SBED biotin
label transfer and analysis by western blotting.

Label transfer involves crosslinking interacting molecules (i.e., bait and prey proteins) with a labeled crosslinking agent and then cleaving the linkage between the bait and prey so that the label remains attached to the prey. This method is particularly valuable because of its ability to identify proteins that interact weakly or transiently with the protein of interest. New non-isotopic reagents and methods continue to make this technique more accessible and simple to perform by any researcher.

Far–western Blot Analysis

Just as pull-down assays differ from co-IP in the detection of protein–protein interactions by using tagged proteins instead of antibodies, so is far–western blot analysis different from western blot analysis, as protein–protein interactions are detected by incubating electrophoresed proteins with a purified, tagged bait protein instead of a target protein-specific antibody, respectively. The term "far" was adopted to emphasize this distinction.

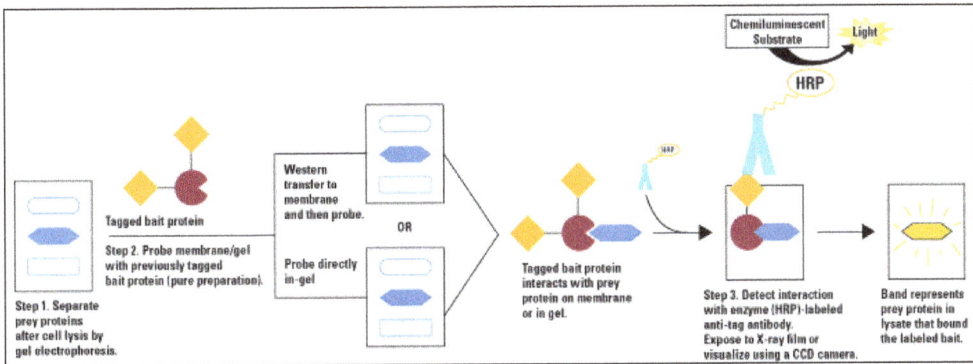

Diagram of far–western blot to analyze protein–protein interactions. In this example, a tagged bait protein is used to probe either the transfer membrane or a gel for the prey protein. Once bound, enzyme (horseradish peroxidase; HRP)-conjugated antibody that targets the bait tag is used to label the interaction, which is then detected by enzymatic chemiluminescence. This general approach can be adjusted by using untagged bait protein that is detected by antibody, biotinylated bait protein that is detected by enzyme-conjugated streptavidin, or radiolabeled bait protein that is detected by exposure to film.

PROTEIN–PROTEIN INTERACTION PREDICTION

Protein–protein interaction prediction is a field combining bioinformatics and structural biology in an attempt to identify and catalog physical interactions between pairs or groups of proteins. Understanding protein–protein interactions is important for the investigation of intracellular signaling pathways, modelling of protein complex structures and for gaining insights into various biochemical processes.

Experimentally, physical interactions between pairs of proteins can be inferred from a variety of techniques, including yeast two-hybrid systems, protein-fragment complementation assays (PCA), affinity purification/mass spectrometry, protein microarrays, fluorescence resonance energy transfer (FRET), and Microscale Thermophoresis (MST). Efforts to experimentally determine the interactome of numerous species are ongoing. Experimentally determined interactions usually provide the basis for *computational methods* to predict interactions, e.g. using homologous protein sequences across species. However, there are also methods that predict interactions *de novo*, without prior knowledge of existing interactions.

Methods

Proteins that interact are more likely to co-evolve, therefore, it is possible to make inferences about interactions between pairs of proteins based on their phylogenetic distances. It has also been observed in some cases that pairs of interacting proteins have fused orthologues in other organisms. In addition, a number of bound protein complexes have been structurally solved and can be used to identify the residues that mediate the interaction so that similar motifs can be located in other organisms.

Phylogenetic Profiling

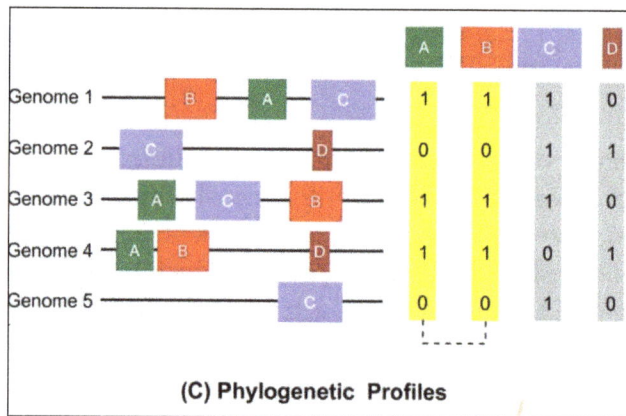

(C) Phylogenetic Profiles

In above the figure, the phylogenetic profiles of four genes (A, B, C and D) are shown on the right. A '1' denotes presence of the gene in the genome and a '0' denotes absence. The two identical profiles of genes A and B are highlighted in yellow.

The phylogenetic profile method is based on the hypothesis that if two or more proteins are concurrently present or absent across several genomes, then they are likely functionally related. *Figure* illustrates a hypothetical situation in which proteins A and B are identified as functionally linked due to their identical phylogenetic profiles across 5 different genomes. The Joint Genome Institute provides an Integrated Microbial Genomes and Microbiomes database (JGI IMG) that has a phylogenetic profiling tool for single genes and gene cassettes.

Prediction of Co-evolved Protein Pairs based on Similar Phylogenetic Trees

It was observed that the phylogenetic trees of ligands and receptors were often more similar than due to random chance. This is likely because they faced similar selection pressures and co-evolved. This method uses the phylogenetic trees of protein pairs to determine if interactions exist. To do this, homologs of the proteins of interest are found (using a sequence search tool such as BLAST) and multiple-sequence alignments are done (with alignment tools such as Clustal) to build distance matrices for each of the proteins of interest. The distance matrices should then be used to build phylogenetic trees. However, comparisons between phylogenetic trees are difficult, and current methods circumvent this by simply comparing distance matrices. The distance matrices of the proteins are used to calculate a correlation coefficient, in which a larger value corresponds to co-evolution. The benefit of comparing distance matrices instead of phylogenetic trees is that the results do not depend on the method of tree building that was used. The downside is that difference matrices are not perfect representations of phylogenetic trees, and inaccuracies may result from using such a shortcut. Another factor worthy of note is that there are background similarities between the phylogenetic trees of any protein, even ones that do not interact. If left unaccounted for, this could lead to a high false-positive rate. For this reason, certain methods construct a background tree using 16S rRNA sequences which they use as the canonical tree of life. The distance matrix constructed from this tree of life is then subtracted from the distance matrices of the proteins of interest. However, because RNA distance matrices and DNA distance matrices have different scale, presumably because RNA and DNA have different mutation rates, the RNA matrix needs to be rescaled before it can be subtracted from the DNA matrices. By using molecular clock proteins, the scaling coefficient for protein distance/RNA distance can be calculated. This coefficient is used to rescale the RNA matrix.

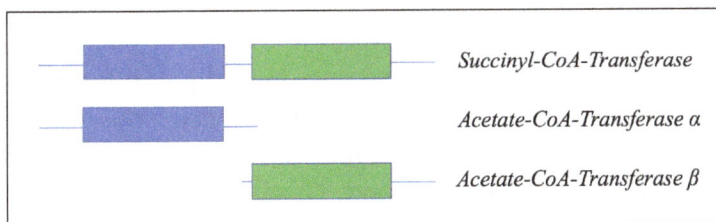

Succinyl-CoA-Transferase

Acetate-CoA-Transferase α

Acetate-CoA-Transferase β

In above the figure, the Human succinyl-CoA-Transferase enzyme is represented by the two joint blue and green bars at the top of the image. The alpha subunit of the Acetate-CoA-Transferase enzyme is homologous with the first half of the enzyme, represents by the blue bar. The beta subunit of the Acetate-CoA-Transferase enzyme is homologous with the second half of the enzyme, represents by the green bar. This mage was adapted from Uetz, P. & Pohl, E. *Protein-Protein and Protein-DNA Interactions.* In: Wink, M. (ed.), Introduction to Molecular Biotechnology, 3rd ed. Wiley-VCH, *in press*.

Rosetta Stone (Gene Fusion) Method

The Rosetta Stone or Domain Fusion method is based on the hypothesis that interacting proteins are sometimes fused into a single protein. For instance, two or more separate proteins in a genome may be identified as fused into one single protein in another genome. The separate proteins are likely to interact and thus are likely functionally related. An example of this is the *Human Succinyl coA Transferase* enzyme, which is found as one protein in humans but as two separate proteins, *Acetate coA Transferase alpha* and *Acetate coA Transferase beta*, in *Escherichia coli*. In order to identify these sequences, a sequence similarity algorithm such as the one used by *BLAST* is necessary. For example, if we had the amino acid sequences of proteins A and B and the amino acid sequences of all proteins in a certain genome, we could check each protein in that genome for non-overlapping regions of sequence similarity to both proteins A and B. *Figure* depicts the BLAST sequence alignment of Succinyl coA Transferase with its two separate homologs in E. coli. The two subunits have non-overlapping regions of sequence similarity with the human protein, indicated by the pink regions, with the alpha subunit similar to the first half of the protein and the beta similar to the second half. One limit of this method is that not all proteins that interact can be found fused in another genome, and therefore cannot be identified by this method. On the other hand, the fusion of two proteins does not necessitate that they physically interact. For instance, the SH2 and SH3 domains in the src protein are known to interact. However, many proteins possess homologs of these domains and they do not all interact.

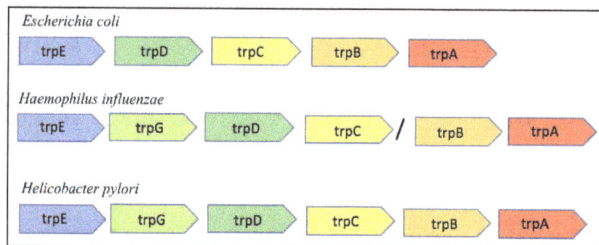

In above the figure, Organization of the trp operon in three different species of bacteria: *Escherichia coli, Haemophilus influenzae, Helicobacter pylori*. Only the trpA and trpB genes are adjacent across all three organisms and are thus predicted to interact by the conserved gene neighborhood method. This image was adapted from Dandekar, T., Snel, B., Huynen, M., & Bork, P. (1998). Conservation of gene order: a fingerprint of proteins that physically interact. *Trends in biochemical sciences*.

Conserved Gene Neighborhood

The conserved neighborhood method is based on the hypothesis that if genes encoding two proteins are neighbors on a chromosome in many genomes, then they are likely functionally related. The method is based on an observation by Bork et al. of gene pair conservation across nine bacterial and archaeal genomes. The method is most effective in prokaryotes with operons as the organization of genes in an operon is generally related to function. For instance, the *trpA* and *trpB* genes in *Escherichia coli* encode the two subunits of the *tryptophan synthase* enzyme known to interact to catalyze a single reaction. The adjacency of these two genes was shown to be conserved across nine different bacterial and archaeal genomes.

Classification Methods

Classification methods use data to train a program (classifier) to distinguish positive examples of interacting protein/domain pairs with negative examples of non-interacting pairs. Popular classifiers used are Random Forest Decision (RFD) and Support Vector Machines. RFD produces results based on the domain composition of interacting and non-interacting protein pairs. When given a protein pair to classify, RFD first creates a representation of the protein pair in a vector. The vector contains all the domain types used to train RFD, and for each domain type the vector also contains a value of 0, 1, or 2. If the protein pair does not contain a certain domain, then the value for that domain is 0. If one of the proteins of the pair contains the domain, then the value is 1. If both proteins contain the domain, then the value is 2. Using training data, RFD constructs a decision forest, consisting of many decision trees. Each decision tree evaluates several domains, and based on the presence or absence of interactions in these domains, makes a decision as to if the protein pair interacts. The vector representation of the protein pair is evaluated by each tree to determine if they are an interacting pair or a non-interacting pair. The forest tallies up all the input from the trees to come up with a final decision. The strength of this method is that it does not assume that domains interact independent of each other. This makes it so that multiple domains in proteins can be used in the prediction. This is a big step up from previous methods which could only predict based on a single domain pair. The limitation of this method is that it relies on the training dataset to produce results. Thus, usage of different training datasets could influence the results.

Inference of Interactions from Homologous Structures

This group of methods makes use of known protein complex structures to predict and structurally model interactions between query protein sequences. The prediction process generally starts by employing a sequence based method (e.g. Interolog) to search for protein complex structures that are homologous to the query sequences. These known complex structures are then used as templates to structurally model the interaction between query sequences. This method has the advantage of not only inferring protein interactions but also suggests models of how proteins interact structurally, which can provide some insights into the atomic level mechanism of that interaction. On the other hand, the ability for these methods to make a prediction is constrained by a limited number of known protein complex structures.

Association Methods

Association methods look for characteristic sequences or motifs that can help distinguish between interacting and non-interacting pairs. A classifier is trained by looking for sequence-signature pairs where one protein contains one sequence-signature, and its interacting partner contains another sequence-signature. They look specifically for sequence-signatures that are found together more often than by chance. This uses a log-odds score which is computed as log2(Pij/PiPj), where Pij is the observed frequency of domains i and j occurring in one protein pair; Pi and Pj are the background frequencies of domains i and j in the data. Predicted domain interactions are those with positive log-odds scores and also having several occurrences within the database. The downside with this method is that it looks at each pair of interacting domains separately, and it assumes that they interact independently of each other.

Identification of Structural Patterns

This method builds a library of known protein–protein interfaces from the PDB, where the interfaces are defined as pairs of polypeptide fragments that are below a threshold slightly larger than the Van der Waals radius of the atoms involved. The sequences in the library are then clustered based on structural alignment and redundant sequences are eliminated. The residues that have a high (generally >50%) level of frequency for a given position are considered hotspots. This library is then used to identify potential interactions between pairs of targets, providing that they have a known structure (i.e. present in the PDB).

Bayesian Network Modelling

Bayesian methods integrate data from a wide variety of sources, including both experimental results and prior computational predictions, and use these features to assess the likelihood that a particular potential protein interaction is a true positive result. These methods are useful because experimental procedures, particularly the yeast two-hybrid experiments, are extremely noisy and produce many false positives, while the previously mentioned computational methods can only provide circumstantial evidence that a particular pair of proteins might interact.

Domain-pair Exclusion Analysis

The domain-pair exclusion analysis detects specific domain interactions that are hard to detect using Bayesian methods. Bayesian methods are good at detecting nonspecific promiscuous interactions and not very good at detecting rare specific interactions. The domain-pair exclusion analysis method calculates an E-score which measures if two domains interact. It is calculated as log(probability that the two proteins interact given that the domains interact/probability that the two proteins interact given that the domains don't interact). The probabilities required in the formula are calculated using an Expectation Maximization procedure, which is a method for estimating parameters in statistical models. High E-scores indicate that the two domains are likely to interact, while low scores indicate that other domains form the protein pair are more likely to be responsible for the interaction. The drawback with this method is that it does not take into account false positives and false negatives in the experimental data.

Supervised Learning Problem

The problem of PPI prediction can be framed as a supervised learning problem. In this paradigm the known protein interactions supervise the estimation of a function that can predict whether an interaction exists or not between two proteins given data about the proteins (e.g., expression levels of each gene in different experimental conditions, location information, phylogenetic profile, etc.).

Relationship to Docking Methods

The field of protein–protein interaction prediction is closely related to the field of protein–protein docking, which attempts to use geometric and steric considerations to fit two proteins of known structure into a bound complex. This is a useful mode of inquiry in cases where both proteins in the pair have known structures and are known (or at least strongly suspected) to interact, but since so many proteins do not have experimentally determined structures, sequence-based interaction prediction methods are especially useful in conjunction with experimental studies of an organism's interactome.

PROTEIN–PROTEIN INTERACTION SCREENING

Protein–protein interaction screening refers to the identification of Protein–protein interaction with high-throughput screening methods such as computer- and robot-assisted plate reading, flow cytometry analyzing.

The interactions between proteins are central to virtually every process in a living cell. Information about these interactions improves understanding of diseases and can provide the basis for new therapeutic approaches.

Methods to Screen Protein–protein Interactions

Though there are many methods to detect protein–protein interactions, the majority of these methods—such as co-immunoprecipitation, fluorescence resonance energy transfer (FRET) and dual polarisation interferometry—are not screening approaches.

Ex vivo or in Vivo Methods

Methods that screen protein–protein interactions in the living cells. Bimolecular fluorescence complementation (BiFC) is a technique for observing the interactions of proteins. Combining it with other new techniques DERB can enable the screening of protein–protein interactions and their modulators.

The yeast two-hybrid screen investigates the interaction between artificial fusion proteins inside the nucleus of yeast. This approach can identify the binding partners of a protein without bias. However, the method has a notoriously high false-positive rate, which makes it necessary to verify the identified interactions by co-immunoprecipitation.

In-vitro Methods

The tandem affinity purification (TAP) method allows the high-throughput identification of proteins interactions. In contrast with the Y2H approach, the accuracy of the method can be compared to those of small-scale experiments and the interactions are detected within the correct cellular environment as by co-immunoprecipitation. However, the TAP tag method requires two successive steps of protein purification, and thus can not readily detect transient protein–protein interactions. Recent genome-wide TAP experiments were performed by Krogan et al., 2006, and Gavin et al., 2006, providing updated protein interaction data for yeast organisms.

Chemical crosslinking is often used to "fix" protein interactions in place before trying to isolate/identify interacting proteins. Common crosslinkers for this application include the non-cleavable [NHS-ester] crosslinker, [bis-sulfosuccinimidyl suberate] (BS3); a cleavable version of BS3, [dithiobis(sulfosuccinimidyl propionate)](DTSSP); and the [imidoester] crosslinker [dimethyl dithiobispropionimidate] (DTBP) that is popular for fixing interactions in ChIP assays.

PHAGE DISPLAY

Phage display is a laboratory technique for the study of protein–protein, protein–peptide, and protein–DNA interactions that uses bacteriophages (viruses that infect bacteria) to connect proteins with the genetic information that encodes them. In this technique, a gene encoding a protein of interest is inserted into a phage coat protein gene, causing the phage to "display" the protein on its outside while containing the gene for the protein on its inside, resulting in a connection between genotype and phenotype. These displaying phages can then be screened against other proteins, peptides or DNA sequences, in order to detect interaction between the displayed protein and those other molecules. In this way, large libraries of proteins can be screened and amplified in a process called *in vitro* selection, which is analogous to natural selection.

The most common bacteriophages used in phage display are M13 and fd filamentous phage, though T4, T7, and λ phage have also been used.

Principle

Like the two-hybrid system, phage display is used for the high-throughput screening of protein interactions. In the case of M13 filamentous phage display, the DNA encoding the protein or peptide of interest is ligated into the pIII or pVIII gene, encoding either the minor or major coat protein, respectively. Multiple cloning sites are sometimes used to ensure that the fragments are inserted in all three possible reading frames so that the cDNA fragment is translated in the proper frame. The phage gene and insert DNA hybrid is then inserted (a process known as "transduction") into *E. coli* bacterial cells such as TG1, SS320, ER2738, or XL1-Blue *E. coli*. If a "phagemid" vector is used (a simplified display construct vector) phage particles will not be released from the *E. coli* cells until they are infected with helper phage, which enables packaging of the phage DNA and assembly of the mature virions with the relevant protein fragment as part of their outer coat on either the minor (pIII) or major (pVIII) coat protein. By immobilizing a relevant DNA or protein target(s) to the surface of a microtiter plate well, a phage that displays a protein that binds to one of those targets on its surface will remain while others are removed by washing. Those that remain can be eluted, used to produce more phage (by bacterial infection with helper phage) and to produce a phage mixture that is enriched with relevant (i.e. binding) phage. The repeated cycling of these steps is referred to as 'panning', in reference to the enrichment of a sample of gold by removing undesirable materials. Phage eluted in the final step can be used to infect a suitable bacterial host, from which the phagemids can be collected and the relevant DNA sequence excised and sequenced to identify the relevant, interacting proteins or protein fragments.

The use of a helper phage can be eliminated by using 'bacterial packaging cell line' technology.

Elution can be done combining low-pH elution buffer with sonification, which, in addition to loosening the peptide-target interaction, also serves to detach the target molecule from the immobilization surface. This ultrasound-based method enables single-step selection of a high-affinity peptide.

Applications

Applications of phage display technology include determination of interaction partners of a protein (which would be used as the immobilised phage "bait" with a DNA library consisting of all coding sequences of a cell, tissue or organism) so that the function or the mechanism of the function of that protein may be determined. Phage display is also a widely used method for *in vitro* protein evolution (also called protein engineering). As such, phage display is a useful tool in drug discovery. It is used for finding new ligands (enzyme inhibitors, receptor agonists and antagonists) to target proteins. The technique is also used to determine tumour antigens (for use in diagnosis and therapeutic targeting) and in searching for protein-DNA interactions using specially-constructed

DNA libraries with randomised segments. Recently, phage display has also been used in the context of cancer treatments - such as the adoptive cell transfer approach. In these cases, phage display is used to create and select synthetic antibodies that target tumour surface proteins. These are made into synthetic receptors for T-Cells collected from the patient that are used to combat the disease.

Competing methods for *in vitro* protein evolution include yeast display, bacterial display, ribosome display, and mRNA display.

Antibody Maturation in Vitro

The invention of antibody phage display revolutionised antibody drug discovery. Initial work was done by laboratories at the MRC Laboratory of Molecular Biology (Greg Winter and John McCafferty), the Scripps Research Institute (Richard Lerner and Carlos F. Barbas) and the German Cancer Research Centre (Frank Breitling and Stefan Dübel). In 1991, The Scripps group reported the first display and selection of human antibodies on phage. This initial study described the rapid isolation of human antibody Fab that bound tetanus toxin and the method was then extended to rapidly clone human anti-HIV-1 antibodies for vaccine design and therapy.

Phage display of antibody libraries has become a powerful method for both studying the immune response as well as a method to rapidly select and evolve human antibodies for therapy. Antibody phage display was later used by Carlos F. Barbas at The Scripps Research Institute to create synthetic human antibody libraries, a principle first patented in 1990 by Breitling and coworkers (Patent CA 2035384), thereby allowing human antibodies to be created in vitro from synthetic diversity elements.

Antibody libraries displaying millions of different antibodies on phage are often used in the pharmaceutical industry to isolate highly specific therapeutic antibody leads, for development into antibody drugs primarily as anti-cancer or anti-inflammatory therapeutics. One of the most successful was adalimumab, discovered by Cambridge Antibody Technology as D2E7 and developed and marketed by Abbott Laboratories. Adalimumab, an antibody to TNF alpha, was the world's first fully human antibody, which achieved annual sales exceeding $1bn.

General Protocol

Below is the sequence of events that are followed in phage display screening to identify polypeptides that bind with high affinity to desired target protein or DNA sequence:

1. Target proteins or DNA sequences are immobilized to the wells of a microtiter plate.

2. Many genetic sequences are expressed in a bacteriophage library in the form of fusions with the bacteriophage coat protein, so that they are displayed on the

surface of the viral particle. The protein displayed corresponds to the genetic sequence within the phage.

3. This phage-display library is added to the dish and after allowing the phage time to bind, the dish is washed.

4. Phage-displaying proteins that interact with the target molecules remain attached to the dish, while all others are washed away.

5. Attached phage may be eluted and used to create more phage by infection of suitable bacterial hosts. The new phage constitutes an enriched mixture, containing considerably less irrelevant phage (i.e. non-binding) than were present in the initial mixture.

6. Steps 3 to 5 are optionally repeated one or more times, further enriching the phage library in binding proteins.

7. Following further bacterial-based amplification, the DNA within in the interacting phage is sequenced to identify the interacting proteins or protein fragments.

Selection of the Coat protein

pIII

pIII is the protein that determines the infectivity of the virion. pIII is composed of three domains (N1, N2 and CT) connected by glycine-rich linkers. The N2 domain binds to the F pilus during virion infection freeing the N1 domain which then interacts with a TolA protein on the surface of the bacterium. Insertions within this protein are usually added in position 249 (within a linker region between CT and N2), position 198 (within the N2 domain) and at the N-terminus (inserted between the N-terminal secretion sequence and the N-terminus of pIII). However, when using the BamHI site located at position 198 one must be careful of the unpaired Cysteine residue (C201) that could cause problems during phage display if one is using a non-truncated version of pIII.

An advantage of using pIII rather than pVIII is that pIII allows for monovalent display when using a phagemid (Ff-phage derived plasmid) combined with a helper phage. Moreover, pIII allows for the insertion of larger protein sequences (>100 amino acids) and is more tolerant to it than pVIII. However, using pIII as the fusion partner can lead to a decrease in phage infectivity leading to problems such as selection bias caused by difference in phage growth rate or even worse, the phage's inability to infect its host. Loss of phage infectivity can be avoided by using a phagemid plasmid and a helper phage so that the resultant phage contains both wild type and fusion pIII.

cDNA has also been analyzed using pIII via a two complementary leucine zippers system, Direct Interaction Rescue or by adding an 8-10 amino acid linker between the cDNA and pIII at the C-terminus.

pVIII

pVIII is the main coat protein of Ff phages. Peptides are usually fused to the N-terminus of pVIII. Usually peptides that can be fused to pVIII are 6-8 amino acids long. The size restriction seems to have less to do with structural impediment caused by the added section and more to do with the size exclusion caused by pIV during coat protein export. Since there are around 2700 copies of the protein on a typical phages, it is more likely that the protein of interest will be expressed polyvalently even if a phagemid is used. This makes the use of this protein unfavorable for the discovery of high affinity binding partners.

To overcome the size problem of pVIII, artificial coat proteins have been designed. An example is Weiss and Sidhu's inverted artificial coat protein (ACP) which allows the display of large proteins at the C-terminus. The ACP's could display a protein of 20kDa, however, only at low levels (mostly only monovalently).

pVI

pVI has been widely used for the display of cDNA libraries. The display of cDNA libraries via phage display is an attractive alternative to the yeast-2-hybrid method for the discovery of interacting proteins and peptides due to its high throughput capability. pVI has been used preferentially to pVIII and pIII for the expression of cDNA libraries because one can add the protein of interest to the C-terminus of pVI without greatly affecting pVI's role in phage assembly. This means that the stop codon in the cDNA is no longer an issue. However, phage display of cDNA is always limited by the inability of most prokaryotes in producing post-translational modifications present in eukaryotic cells or by the misfolding of multi-domain proteins.

While pVI has been useful for the analysis of cDNA libraries, pIII and pVIII remain the most utilized coat proteins for phage display.

pVII and pIX

In an experiment in 1995, display of Glutathione S-transferase was attempted on both pVII and pIX and failed. However, phage display of this protein was completed successfully after the addition of a periplasmic signal sequence (pelB or ompA) on the N-terminus. In a recent study, it has been shown that AviTag, FLAG and His could be displayed on pVII without the need of a signal sequence. Then the expression of single chain Fv's (scFv), and single chain T cell receptors (scTCR) were expressed both with and without the signal sequence.

PelB (an amino acid signal sequence that targets the protein to the periplasm where a signal peptidase then cleaves off PelB) improved the phage display level when compared to pVII and pIX fusions without the signal sequence. However, this led to the incorporation of more helper phage genomes rather than phagemid genomes. In all cases, phage display levels were lower than using pIII fusion. However, lower display might be more favorable for the selection of binders due to lower display being closer to true monovalent display. In five out of six occasions, pVII and pIX fusions without pelB was more efficient than pIII fusions in affinity selection assays. The paper even goes on to state that pVII and pIX display platforms may outperform pIII in the long run.

The use of pVII and pIX instead of pIII might also be an advantage because virion rescue may be undertaken without breaking the virion-antigen bond if the pIII used is wild type. Instead, one could cleave in a section between the bead and the antigen to elute. Since the pIII is intact it does not matter whether the antigen remains bound to the phage.

T7 phages

The issue of using Ff phages for phage display is that they require the protein of interest to be translocated across the bacterial inner membrane before they are assembled into the phage. Some proteins cannot undergo this process and therefore cannot be displayed on the surface of Ff phages. In these cases, T7 phage display is used instead. In T7 phage display, the protein to be displayed is attached to the C-terminus of the gene 10 capsid protein of T7.

The disadvantage of using T7 is that the size of the protein that can be expressed on the surface is limited to shorter peptides because large changes to the T7 genome cannot be accommodated like it is in M13 where the phage just makes its coat longer to fit the larger genome within it. However, it can be useful for the production of a large protein library for scFV selection where the scFV is expressed on an M13 phage and the antigens are expressed on the surface of the T7 phage.

Bioinformatics Resources and Tools

Databases and computational tools for mimotopes have been an important part of phage display study. Databases, programs and web servers have been widely used to exclude target-unrelated peptides, characterize small molecules-protein interactions and map protein-protein interactions. Users can use three dimensional structure of a protein and the peptides selected from phage display experiment to map conformational eptiopes. Some of the fast and efficient computational methods are available online.

TANDEM AFFINITY PURIFICATION

Tandem affinity purification (TAP) is an immunoprecipitation-based purification technique for studying protein–protein interactions. The goal is to extract from a cell only the protein of interest, in complex with any other proteins it interacted with. TAP uses two types of agarose beads that bind to the protein of interest and that can be separated from the cell lysate by centrifugation, without disturbing, denaturing or contaminating the involved complexes. To enable the protein of interest to bind to the beads, it is tagged with a designed piece, the TAP tag.

The original TAP method involves the fusion of the TAP tag to the C-terminus of the protein under study. The TAP tag consists of calmodulin binding peptide (CBP) from the N-terminal, followed by a TEV protease cleavage site and two Protein A domains, which bind tightly to IgG (making a TAP tag a type of epitope tag).

Many other tag/bead/eluent combinations have been proposed since the TAP principle was first published.

Variant Tags

This tag is also known as the C-terminal TAP tag because an N-terminal version is also available. However, the method to be described assumes the use of a C-terminal tag, although the principle behind the method is still the same.

Process

There are a few methods in which the fusion protein can be introduced into the host cells. If the host is yeast, then one of the methods may be the use of plasmids that will eventually translate the fusion protein within the host. Whichever method that is being used, it is preferable to maintain expression of the fusion protein as close as possible to its natural level. Once the fusion protein is translated within the host, it will interact with other proteins, ideally in a manner unaffected by the TAP tag.

Subsequently, the tagged protein (with its binding partners) is retrieved using an affinity selection process.

The first type of bead added is coated with Immunoglobulin G, which binds to the TAP tag's outermost end. The beads, with the proteins of interest, are separated from the lysate via centrifugation. The proteins are then released from the beads by an enzyme (TEV protease) which breaks the tag at the TEV cleavage site in the middle.

After this first purification step, a second type of bead (coated with calmodulin) is added to the released proteins which binds reversibly to the remaining piece of the TAP tag still on the proteins. The beads are again separated by centrifugation, further removing contaminants as well as the TEV protease. Finally, the beads are released by EGTA, leaving behind the native eluate containing only the protein of interest, its bound protein partners and the remaining CBP piece of the TAP tag.

The native eluate can then be analyzed using gel electrophoresis and mass spectrometry to identify the protein's binding partners.

Advantages

An advantage of this method is that there can be real determination of protein partners quantitatively in vivo without prior knowledge of complex composition. It is also simple to execute and often provides high yield. One of the obstacles of studying protein protein interaction is the contamination of the target protein especially when we don't have any prior knowledge of it. TAP offers an effective, and highly specific means to purify target protein. After 2 successive affinity purifications, the chance for contaminants to be retained in the eluate reduces significantly.

Disadvantages

However, there is also the possibility that a tag added to a protein might obscure binding of the new protein to its interacting partners. In addition, the tag may also affect protein expression levels. On the other hand, the tag may also not be sufficiently exposed to the affinity beads, hence skewing the results.

There may also be a possibility of a cleavage of the proteins by the TEV protease, although this is unlikely to be frequent given the high specificity of the TEV protease.

Suitability

As this method involves at least 2 rounds of washing, it may not be suitable for screening transient protein interactions, unlike the yeast two-hybrid method or *in vivo* crosslinking with photo-reactive amino acid analogs. However, it is a good method for testing stable protein interactions and allows various degrees of investigation by controlling the number of times the protein complex is purified.

Applications

In 2002, the TAP tag was first used with mass spectrometry in a large-scale approach to systematically analyse the proteomics of yeast by characterizing multiprotein complexes. The study revealed 491 complexes, 257 of them wholly new. The rest were familiar from other research, but now virtually all of them were found to have new components. They drew up a map relating all the protein components functionally in a complex network.

Many other proteomic analyses also involve the use of TAP tag. A research by EMBO (Dziembowski, 2004) identified a new complex required for nuclear pre-mRNA retention and splicing. They have purified a novel trimeric complex composed of 3 other subunits (Snu17p, Bud13p and Pml1p) and find that these subunits are not essential for viability but required for efficient splicing (removal of introns) of pre-mRNA. In 2006, *Fleischer et al.* systematically identified proteins associated with eukaryotic ribosomal complexes. They used multifaceted mass spectrometry proteomic screens to identify yeast ribosomal complexes and then used TAP tagging to functionally link up all these proteins.

Other Epitope-tag Combinations

The principle of tandem-affinity purification of multiprotein complexes is not limited to the combination of CBP and Protein A tags used in the original work by Rigaut et al.. For example, the combination of FLAG- and HA-tags has been used since 2000 by the group of Nakatani to purify numerous protein complexes from mammalian cells. Many other tag combinations have been proposed since the TAP principle was published.

CROSSLINKING FOR PROTEIN INTERACTION ANALYSIS

Crosslinking reagents covalently link together interacting proteins, domains or peptides by forming chemical bonds between specific amino acid functional groups on two or more biomolecules that occur in close proximity because of their interaction. Commercially available crosslinking reagents have a wide range of characteristics, including:

- Functional group specificity: The crosslinker molecule carries reactive moieties that target amines, sulfhydryls, carboxyls, carbonyls or hydroxyls.

- Homobifunctional or heterobifunctional: Molecules are available with identical reactive moieties on both ends (termed homobifunctional molecules; the upper molecule (A/A) shown below, e.g., DSS) that crosslink identical residues, while each reactive group on heterobifunctional crosslinkers targets different functional groups on separate proteins for greater variability or specificity, as shown in the lower molecule below (A/B, e.g., SMCC).

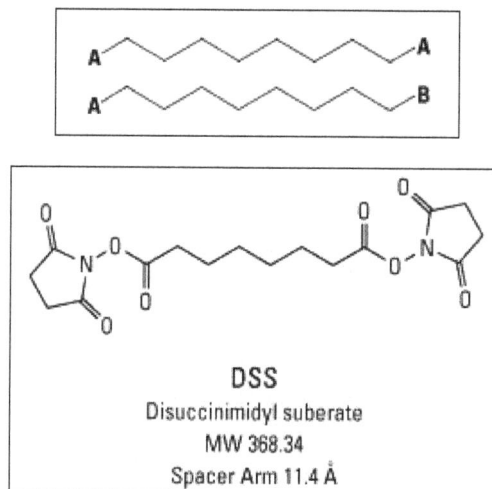

DSS
Disuccinimidyl suberate
MW 368.34
Spacer Arm 11.4 Å

Chemical structure of DSS.

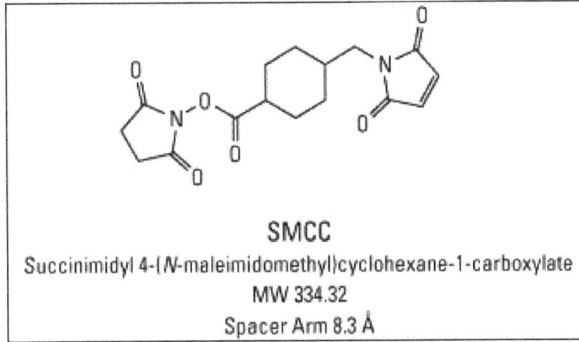

SMCC
Succinimidyl 4-(*N*-maleimidomethyl)cyclohexane-1-carboxylate
MW 334.32
Spacer Arm 8.3 Å

Chemical structure of SMCC.

- Variable spacer arm length: The reactive groups are spatially separated by the crosslinker molecule structure, which allows the crosslinking of amino acids that are varying distances apart, as shown below with Sulfo-EGS versus BS3. Zero-length crosslinkers are also available, which crosslink two amino acid residues without leaving any part of the crosslinker molecule remaining in the interaction after the reaction is completed.

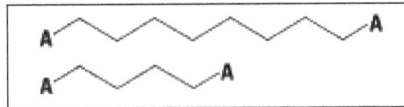

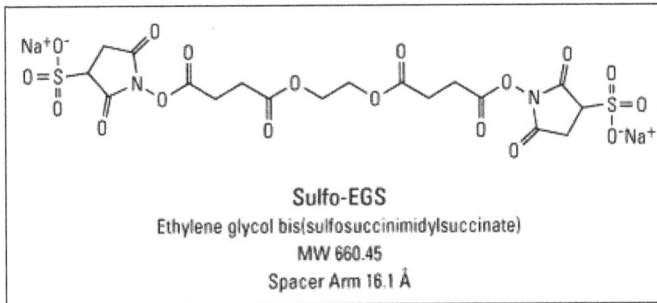

Sulfo-EGS
Ethylene glycol bis(sulfosuccinimidylsuccinate)
MW 660.45
Spacer Arm 16.1 Å

Chemical structure of Sulfo-EGS.

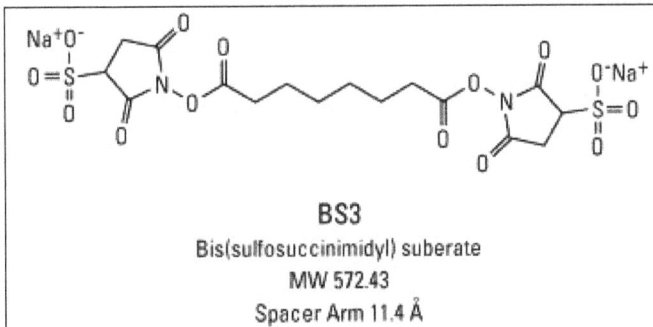

BS3
Bis(sulfosuccinimidyl) suberate
MW 572.43
Spacer Arm 11.4 Å

Chemical structure of BS3.

- Cleavable or non-cleavable: The crosslinker molecule can also be designed to include cleavable elements, such as esters or disulfide bonds (diagrammed below, e.g., DSP), to reverse or break the linkage by the addition of hydroxylamine or reducing agents, respectively.

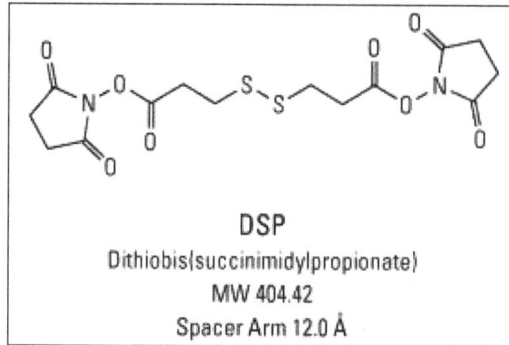

DSP
Dithiobis(succinimidylpropionate)
MW 404.42
Spacer Arm 12.0 Å

Chemical structure of DSP.

- Water-soluble or -insoluble: Crosslinkers can be hydrophobic to allow passage into hydrophobic protein domains (DSS) or through the cell membrane or hydrophilic to limit crosslinking to aqueous compartments (BS3).

In Vivo Crosslinking

Besides the transient and sometimes tentative nature of some protein–protein interactions, the formation of these complexes can change in response to any number of stimuli, including changes in pH, temperature and osmolarity, and either the lack of a specific protein or co-factor or the introduction of a protein with which the proteins do not normally interact.

The benefit of in vivo crosslinking is that the protein–protein interaction can be captured in its native environment, which limits the risk of false positive interactions or the loss of complex stability during cell lysis. For in vivo crosslinking, hydrophobic, lipid-soluble crosslinkers are expected to be used if the target protein is within or across cell membranes, while hydrophilic, water-soluble crosslinkers can be used to crosslink cell surface proteins, such as receptor–ligand complexes. This representative data provides an example of various reagents used for in vivo crosslinking.

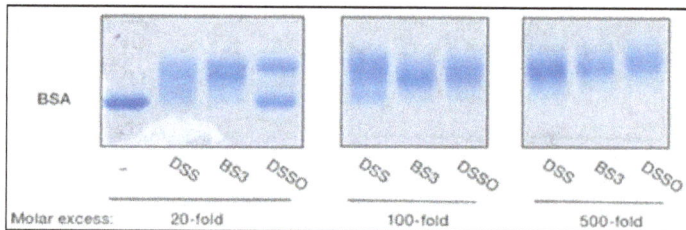

Comparison of several in vivo crosslinking methods. HeLa cells treated with 1% Formaldehyde (HCHO) or 1 mM homobifunctional NHS-ester crosslinker (Thermo Scientific DSG and DSS) in PBS for 10 minutes before quenching. A fourth set of HeLa cells were treated and crosslinked for 10 minutes with 4 mM Photo-Leucine, 2 mM

Photo-Methionine (Photo-AA) according to the procedure. Formaldehyde-treated and NHS-ester–treated cells were quenched with 100 mM glycine (pH 3) and 500 mM Tris (pH 8.0), respectively for an additional 15 minutes. One million cells from each condition were then lysed and 10 µg of each sample was heated at 65°C for 10 minutes in reducing sample buffer containing 50 mM DTT followed by analysis by SDS-PAGE and western blotting with Stat3 specific antibodies (Cell Signaling). Gapdh (Santa Cruz) and beta-actin (US Biologicals) were blotted as loading controls.

Due to the high concentration of proteins in cells, crosslinkers with shorter spacer arms are usually recommended for in vivo crosslinking approaches to increase the specificity of conjugating actual interacting proteins as opposed to proteins that just happen to be in close proximity to each other during incubation with the crosslinker.

Although in vivo crosslinking can yield physiologically relevant, stably-crosslinked complexes for analysis, optimizing this approach can be difficult, as the reaction conditions cannot be tightly controlled and crosslinkers react with a wide array of proteins that all present functional groups against which crosslinkers specifically react.

In Vitro Crosslinking

In vitro crosslinking can better target specific crosslinking events, because more reaction conditions can be tightly controlled, including the pH, temperature, concentration of reactants and purity of the target protein(s). The ability to control all aspects of a conjugation experiment results in better analysis due to greater resolution of protein–protein interactions. Additionally, in vitro methods of conjugation allow researchers to modify interacting proteins, such as adding polyethylene glycol groups (PEGylation), blocking sulfhydryls or converting amines to sulfhydryls. Also, a greater variety of crosslinking reagents, both hydrophobic and hydrophilic, are available for in vitro applications. This representative data was produced using the amine-reactive crosslinking reagents, DSS, BS3, and DSSO.

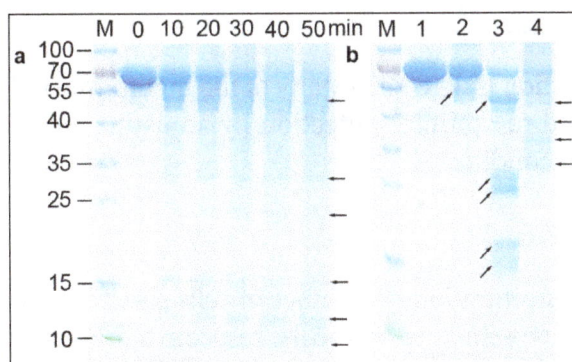

Comparison of BSA crosslinking efficiency by SDS-PAGE.

Different crosslinkers were incubated with BSA at molar excess of crosslinker to protein (e.g., 20-, 100- or 500-fold). Crosslinking efficiency is shown by decreased mobility by SDS-PAGE and varied by crosslinker type, solubility and concentration.

Obviously, the disadvantage of using in vitro methods to conjugate proteins is the lack of physiological conditions. Additionally, rupturing and solubilizing membranes can disrupt protein–protein and protein–membrane interactions.

Because a myriad of crosslinking reagents are commercially available for many different applications, the key determinant in deciding to use in vivo or in vitro crosslinking is the target protein, specifically in term of its:

- Cellular location: In vivo crosslinking would benefit protein targets embedded in the cell membrane, while cytoplasmic proteins could be crosslinked by either method, depending on the next determinant.

- Interaction stability: Weak protein–protein interactions may be lost during in vitro crosslinking due to cell lysis and potential competition with other proteins, while stable interactions may be strong enough to withstand these forces.

General Reaction Conditions

Choosing the Appropriate Crosslinker

Correct identification of protein-protein interactions first requires the selection of the best crosslinker to use. Because there are multiple amino acid functional groups that may react with different crosslinkers, an empirical strategy of screening multiple types of crosslinkers should first be performed to identify the target protein conjugate. The crosslinkers tested may vary in:

- Hydrophobicity.

- Reactive groups.

- Homo- vs. heterobifunctionality.

- Spacer arm length.

Once the target interaction is detected by any of the methods listed below, then the protocol can be fine-tuned to optimize detection by adjusting crosslinker concentration, pH and other reaction conditions.

Sample Preparation

The starting protein concentration or number of cells should be empirically determined for in vitro and in vivo crosslinking protocols, respectively. For in vitro crosslinking, the protein solution should be prepared in a nonreactive buffer, such as phosphate-buffered saline (PBS), which has the proper pH for the specific crosslinker. For in vivo crosslinking applications, cells should be in the exponential phase of growth and at a subconfluent density during the crosslinking procedure. To avoid the possibility of culture media reacting with the crosslinker, the media can be replaced with PBS through a series of cell washes.

Reaction Conditions

Crosslinkers should be prepared as per the manufacturer's instructions; hydrophobic crosslinkers are first dissolved in the appropriate solvent, such as methanol or acetone. The optimum amount of reagent to add also depends on the crosslinker, but usually a 20- to 500-fold molar excess (relative to the lysate protein concentration) is appropriate. Ensure that pH of the reaction buffer is favorable for the crosslinker. Most amine-reactive crosslinkers require alkaline pH for activity.

The crosslinking reaction time may also be important, depending upon the experiment and crosslinker being used. While 30 minutes is a good incubation time to start with, multiple experiments can be performed concurrently to test other lengths of time to determine the optimal time of incubation with the specific crosslinker. Long incubation periods should generally be avoided, not only because it may cause formation of large, crosslinked protein aggregates, but also because the crosslinker may lose stability. In cases where extended incubation periods are required, though, fresh crosslinker can be added at specific time points throughout the procedure to maintain the proper molar ratio of reagent and maximize the formation of the target product. The formation of aggregates due to extensive crosslinking, though, should also be considered in determining the optimal reaction time.

Quenching the Reaction

With most amine-reactive crosslinkers used for protein–protein interaction analysis, the reaction can be halted at the desired time by adding excess nucleophile, such as Tris or glycine, which out-competes the lysate proteins for reaction with the crosslinker. The crosslinked product can then be purified through multiple approaches, including precipitation, chromatography, dialysis or ultrafiltration.

A rapid method that combines quenching the reaction and denaturing the proteins in preparation for gel electrophoresis is to add sodium dodecyl sulfate–polyacrylamide gel electrophoresis (SDS-PAGE) buffer, which contains both Tris and 2-Mercatpoethanol, and then boil the solution for 5 minutes. The sample can then be directly analyzed by gel electrophoresis.

Protein–protein Interaction Analysis

Crosslinking is typically used to capture and stabilize transient or labile interactions so that they can be further isolated and analyzed by downstream methods such as electrophoresis, staining, western blot, immunoprecipitation or co-immunoprecipitation and mass spectrometry.

Western Blot

When two proteins are covalently crosslinked, the gel migration patterns of both proteins shift in relation to the uncrosslinked proteins. Therefore, if antibodies that detect each target protein are available, the most straightforward method to detect the shift of the interacting proteins is by SDS-PAGE and western blot analysis.

Immunoprecipitation and Co-immunopreciptation

Both immunoprecipitation (IP) and co-immunoprecipitation (co-IP) are methods to detect protein expression and protein–protein interactions, respectively, via affinity purification. Crosslinking is commonly performed in both applications, either alone or in combination with affinity binding, to immobilize antibody to the beaded support and or freeze weak antibody–antigen interactions to prevent sample loss during immune complex extraction. Crosslinking is also used to stabilize transient or weak protein–protein interactions prior to co-IP protocols. Following both approaches, samples are commonly analyzed by SDS-PAGE.

Mass Spectrometry

When analysis by mass spectrometry (MS) is available, the peptide fragments that are crosslinked between interacting proteins can be identified by the change in mass resulting from the attached crosslinker molecule. In this approach, identical samples are crosslinked with either deuterated (heavy) or nondeuterated (light) crosslinkers. The crosslinked proteins are then pooled together and analyzed by MS to identify and quantify the heavy product based on its shift from the light product. This method also commonly employs SDS-PAGE as a first-stage purification step prior to digestion in preparation for MS analysis.

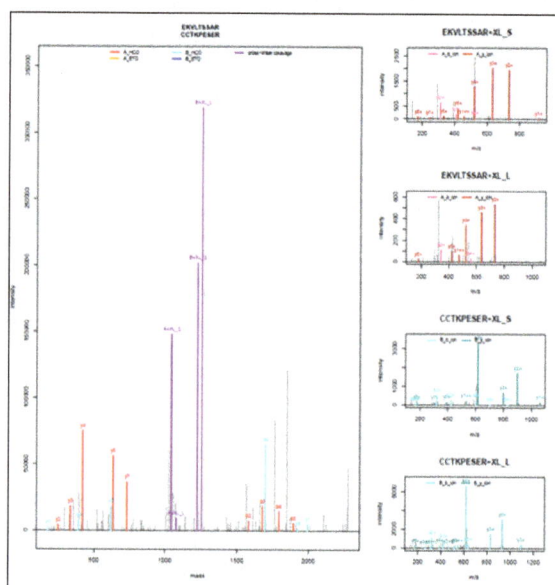

BSA crosslinked peptide spectra.

BSA crosslinked peptide spectra were identified by MS2-MS3 method and XLinkX using DSSO crosslinker. XlinkX software uses unique fragment ion patterns of MS-cleavable crosslinkers (purple annotation) to detect and filter crosslinked peptides for a crosslinked database search.

References

- Shoemaker, BA; Panchenko, AR (2007). "Deciphering protein–protein interactions. Part II. Computational methods to predict protein and domain interaction partners". Plos Comput Biol. 3 (4): e43. Bibcode:2007PLSCB...3...43S. Doi:10.1371/journal.pcbi.0030043. PMC 1857810. PMID 17465672

- Pierce-protein-methods/overview-protein-protein-interaction-analysis, pierce-protein-methods, protein-biology-resource-library, protein-biology-learning-center, protein-biology, life-science, home: thermofisher.com, Retrieved 29 August, 2019

- Scott JS, Barbas CF III, Burton DA (2001). Phage Display: A Laboratory Manual. Plainview, N.Y: Cold Spring Harbor Laboratory Press. ISBN 978-0-87969-740-2

- Crosslinking-protein-interaction-analysis, pierce-protein-methods, protein-biology-resource-library, protein-biology-learning-center, protein-biology, life-science, home: thermofisher.com, Retrieved 30 January, 2019

- Bouveret E, et al. (2000). "A Sm-like protein complex that participates in mrna degradation". The EMBO Journal. 19 (7): 1661–1671. Doi:10.1093/emboj/19.7.1661. PMC 310234. PMID 10747033

PERMISSIONS

INDEX

www.ingramcontent.com/pod-product-compliance
Lightning Source LLC
Chambersburg PA
CBHW061939190326
41458CB00009B/2785